58 Topics in Current Chemistry

Fortschritte der chemischen Forschung

# New Theoretical Aspects



Springer-Verlag Berlin Heidelberg GmbH 1975

This series presents critical reviews of the present position and future trends in modern chemical research. It is addressed to all research and industrial chemists who wish to keep abreast of advances in their subject.

As a rule, contributions are specially commissioned. The editors and publishers will, however, always be pleased to receive suggestions and supplementary information. Papers are accepted for "Topics in Current Chemistry" in either German or English.

ISBN 978-3-662-15863-0 ISBN 978-3-540-37570-8 (eBook)
DOI 10.1007/978-3-540-37570-8

Library of Congress Cataloging in Publication Data. Main entry under title: The theoretical aspects. (Topics in current chemistry; 58). Bibliography: p. Includes index. CONTENTS: Julg, A. On the description of molecules using point charges and electric moments. – Butler, R. S. and DeMaine P. A. D. CRAMS, an automatic chemical reaction analysis and modeling system. — Geick, R. IR Fourier transform spectroscopy. 1. Chemistry — Addresses, essays, lectures. I. Julg, André, 1926 — II. Series.
QDI.F58 Bd. 58 [QD39] 540 .8s [540] 75-17852


Originally published by Springer-Verlag Berlin Heidelberg New York in 1975.
Softcover reprint of the hardcover 1st edition 1975

## Contents

| | |
|---|---|
| Springer-Verlag | D-6900 Heidelberg 1 · Postfach 105280<br>Telephone (06221) 487-1 · Telex 04-61723<br>D-1000 Berlin 33 · Heidelberger Platz 3<br>Telephone (030) 822001 · Telex 01-83319 |
| Springer-Verlag New York Inc. | New York, NY 10010 · 175, Fifth Avenue<br>Telephone 673-2660 |

# On the Description of Molecules Using Point Charges and Electric Moments

**Prof. André Julg**

Laboratoire de Chimie Théorique, Université de Provence, Marseille, France

## Contents

## I. Introduction

The idea of electrical charge in molecules is a very old one. Since it has always been connected with the general theory of molecules and of chemical reactions, its history reflects the same changes in perspective.

As a result of his discovery of electrolysis, Davy built a theory around the idea that chemical combination resulted from electrostatic attraction between electrically charged particles, the electrical charge appearing when the particles came together [1]. Berzelius developed this idea in 1819 and proposed a general theory of reactions, according to which the charges were already present in atoms [2] and every atom had both electricities, positive and negative, more or less concentrated around the poles. This led him to classify the elements according to their prevailing electrical state, positive or negative: oxygen, chlorine, etc. were negative; hydrogen, the metals, etc. were positive.

This theory was strongly contested by Dumas and his school, who showed in 1834 that a chlorine atom, which should have been negative according to Berzelius, could replace a (positive) hydrogen atom in molecules as acetic acid or the hydrogen carbides [3]. A great controversy followed, the chemists being divided into two camps: those interested in the so-called mineral molecules, partisans of the electrical theory of Berzelius, and those who studied the so-called organic molecules, for which Dumas theory of substitution seemed necessary.

This unsatisfactory situation lasted till the end of the nineteenth century, when the electron was discovered [4] and the structure of atoms clarified. Now chemistry entered a new era. The old intuitive idea of affinity between atoms was replaced by the idea of affinity of atoms for electrons. Each atom was characterized by its attractive or repulsive power for electrons [5]. It followed that in a molecule, if two atoms had a different attractive power, the binding between these two atoms was polar, even if the molecule was not completely ionized [6]. At the same time the idea was slowly forced upon chemists by experience that all reactions must be preceded by ionic dissociation of molecules, the polar bonds dissociating of course more easily than the non-polar bonds [7].

Finally, about 1920, Lewis explicitly attributed the formation of bonds to electron pairs, and thus completed the general description of molecules [8]; he introduced the distinction between homopolar or covalent and ionic bonds.

This way of viewing molecular structure and chemical reactions bridged the two opposing theories of Berzelius and Dumas. Its success was extraordinary, as is well known [9]. For instance, hydrolysis of methyliodide was explained by dissociation of water in ions $H^+$ and $OH^-$, followed by replacement of iodine (negative) by the $OH^-$ ions:

$$C_2H_5^+I^- + H^+OH^- \longrightarrow C_2H_5OH + HI$$

But this theory proved useful mainly in the field of addition reactions: for example, for the carbonyl group, the sites of addition of asymmetric reactants could be predicted from the signs of the ions into which they dissociated and from the polarities, negative for oxygen and positive for carbon, which resulted from their different power of attraction for electrons. The double bond of CO took the ionic form $C^+O^-$:

$$>C = O + HCN \longrightarrow >C^+ \begin{matrix} \\ \\ O^- \end{matrix} + \begin{matrix} CN^- \\ H^+ \end{matrix} \longrightarrow >C \begin{matrix} CN \\ \\ OH \end{matrix}$$

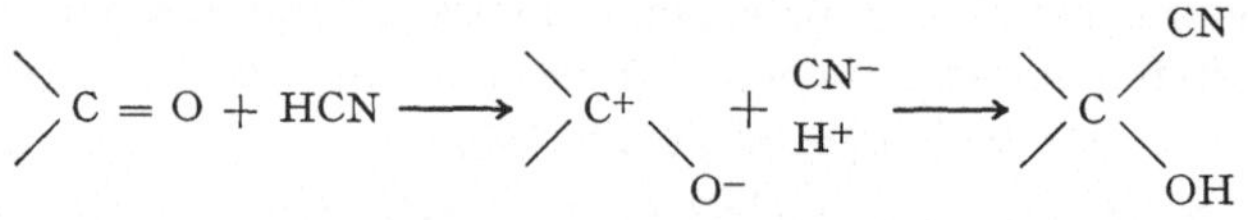

Experiments showed on the other hand that a hydracid molecule could be fixed by an asymmetrically substituted C=C double bond with a well defined orientation (Markovnikow's rule) [10]. Hydrogen was directed to the more substituted carbon atom. This fact was explained by the polarization

$$\begin{matrix} R \\ \\ H \end{matrix} >C^+ - C^- < \begin{matrix} H \\ \\ H \end{matrix}$$

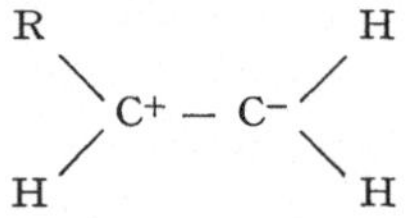

of the double bond. By analogy with the carbonyl group, it was possible to conclude that the most substituted carbon atom was the one least rich in electrons.

In the same way the orientation rules for the second substitution on the benzene ring [11] found a simple explanation if alternant polarity on the ring was assumed [12] to result from the electron-attracting or -repelling power of the first substituent. For example,

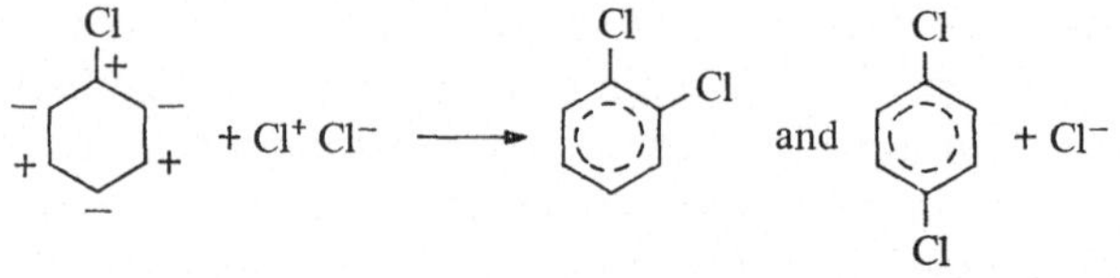

A fundamental remark must be made at this point. If it is so that charges direct an electrically charged reactant towards certain points of a molecule when an ionic process takes place, the approach itself of a reactant changes the initial charge distribution in the molecule because of the electrical field it creates. This is the polarization phenomenon. It follows, for instance, in the case of the carbonyl group mentioned above, that if the negative charge situated in the neighborhood of the oxygen nucleus attracts the $H^+$ ion, and the positive charge of carbon attracts the $CN^-$ ion, without the effect of the field created by this ion, the negative charge of oxygen will decrease to complete ionization ($C^+ - O^-$) when at least one of the two ions $H^+/CN^-$ is sufficiently close. If the charges present in the isolated molecule are sufficiently large, their value will determine the point of attack; this is produced for instance for the methylation of azulene by $ICH_3$. The $CH_3^+$ ion will come on carbon 1 (or 3) [13] which is strongly negatively charged [14]. Under the action of the field created by $CH_3^+$, a $-1$ charge will appear at that point. On the other hand, in a molecule such as naphtalene, where the net charges are very weak [15], the point of attack of a positive ion ($SO_3H^+$ for instance) will be conditioned by the ease with which a $+1$ charge appears on the $\alpha$ position [16].

Another example is provided by pyrrole [17]. When an ionic process takes place, the charges are clearly not the only factor. In addition, it is well known today that several reactions exist which do not take place according to an ionic mechanism: the Kharash effect [18], *i.e.* the fixation of a molecule of hydracid on a double C=C bond in the opposite sense of that predicted by the Markovnikow rule, the substitution reactions by chlorine in saturated hydrocarbons, etc. Nevertheless, the notion of electrical charges present in the isolated molecules retains a great interest for the chemist in interpreting a large number of reactions.

It is also interesting to note that, parallel to the researches of chemists, physicists also have been interested in the problem of charges in molecules, but in a global form, and had in fact succeeded in obtaining important results.

Considering a molecule as equivalent from the electrostatical point of view to a dipole, which could be polarized under the effect created by its neighbors, Mosotti and Clausius, and independently from each other Lorentz and Lorenz [19], obtained the general connection between the dielectric constant of a substance in the gas state and its dipole moment:

$$\frac{4\pi N}{9kT}\mu^2 = \left(\frac{\varepsilon-1}{\varepsilon+2} - \frac{n^2-1}{n^2+2}\right)\frac{M}{d}$$

Later, Onsager generalized this equation by extending it to the case of liquids [20]. As a result of the work of Debye who in 1912 formulated the principle of experimental determination of electric dipole moments [21], several results were obtained. Nevertheless, the chemists had no way of extracting the values of the electrical charges whose association was suggested by the existence of a dipole moment.

Parallel to these studies, the study of real gases had led to the introduction of interaction forces between molecules, responsible for the differences between real fluids and perfect gases. These forces could be attributed to an electrostatic origin: dipole force, polarization and orientation effects [22].

Thus, the non-uniform repartition of positive and negative charges in a molecule was a well-established fact. The crucial problem remained of determining their exact distribution. Two important results were nevertheless obtained: one quantitative result, the value of the dipole moment, the other only a qualitative but equally important result, the signs and the relative order of charges situated in the neighborhoods of different nuclei. To go beyond that, it was necessary to have recourse to a theory capable of successfully attacking the problem of molecular structure itself. This was the role of quantum mechanics.

## II. Description of Molecules According to Quantum Mechanics

### A. Quantum Theory

Without entering into details, it is necessary to recall the general characteristic methods of Quantum Mechanics in order clearly to understand the problem to be solved [23].

The idea of trajectory in the classical sense of particles forming the system is not introduced. All the information available on the motion of particles is reduced to knowledge of a function of the coordinates of the particles and of time—called wave function associated to the system: $\Psi(q,t)$. The square of the absolute value of this function, $|\Psi|^2$, represents the probability of finding the particles in the points of coordinates $q$, at the time $t$. In an isolated system with a well defined energy, this probability does not depend on time, but only on the coordinates $q$. The system is then declared to be in a stationary state.

Our experiments refer necessarily to a large number of systems, atoms or molecules. As far as they can be considered as independent systems, *i.e.* as having the same characteristics, the result of a measurement made at the macroscopic scale will give the average of the results which would be obtained by carrying out a very large number of microscopic measurements after a single system. It follows that in a molecule the square of the modulus of the wave function associated with the electrons will be interpreted as giving the density of a continuous repartition of negative charge. This is what we call the "electron cloud". The dynamic model is replaced by a static model.

The wave equation $\Psi$ is a solution of the partial differential equation of Schrödinger [24]. The construction of this equation makes use of a mathematical formalism which we shall not treat here.

In principle, owing to its generality, the Schrödinger equation contains the solution of all problems of molecular structure. Unfortunately, it is only integrable in a few special cases, the hydrogen atom for instance.

Thus, in general, one must be content with approximate solutions.

## B. Application to Molecules

The specific case of molecules lends itself to a first important simplification. In fact, the very large difference between nuclei and electrons allows reduction of the problem to electrons only. More precisely, the wave function can be determined to describe the electrons for every geometrical configuration of the nuclei. This is the Born-Oppenheimer approximation [25]. The knowledge of the purely electronic wave function allows calculation of the total molecular energy including the repulsion energy of the positive nuclei. In general this energy is at a minimum for a particular arrangement of the nuclei. This is the equilibrium geometry around which the molecule will vibrate. Neglecting vibrations, we shall assume in the following that the nuclei are fixed in their equilibrium positions.

The study of the $H_2$ molecule [26] showed that, in agreement with the prophetical ideas of Lewis [8], the two electrons of that molecule do in fact pair together and the resulting electron density is concentrated between the nuclei. Generalizing this result, it is reasonable to try to describe in general a molecule in terms of electron pairs. In the $H_2$ molecule, with a good approximation, one can describe each of the two electrons by the same spatial function, *i.e.* assign to each electron the same spatial distribution of the charge. The energy obtained under this assumption is actually little different from the exact value [27]. By extrapolation, in a general molecule, each electron of every pair will be assigned a space function, the *molecular orbital*, the two electrons of a given pair being described by the same space function (they will differ only by their spins).

This simplifying assumption allows construction of a spatial wave function describing all the electrons of a molecule, having an even number of electrons, in the ground state, in the form of a determinant [28]:

$$\psi(1, 2, \ldots, n) = \frac{1}{\sqrt{n!}} \begin{vmatrix} \varphi_1(1)\,\alpha(1) & \varphi_1(1)\,\beta(1) & \varphi_2(1)\,\alpha(1)\ldots \\ \varphi_1(2)\,\alpha(2) & \varphi_1(2)\,\beta(2) & \varphi_2(2)\,\alpha(2)\ldots \\ \ldots & \ldots & \ldots \end{vmatrix} \tag{1}$$

Here, $n$ is the number of electrons, the $\varphi_i$'s are the space functions, and $\alpha$ and $\beta$ are the functions associated with the electron spin. (The choice of the structure of the determinant is imposed by a general theorem of quantum mechanics, which says that the wave function must change its sign when two electrons are interchanged). We leave aside the case of molecules whose levels are not doubly occupied, viz. radicals [29].

The problem is thus reconducted to the determination of the various functions $\varphi_i$. A well-known procedure consists in expanding these functions in a basis of conveniently chosen functions [30]:

$$\varphi_i = \sum_r c_{ir}\,\chi_r \tag{2}$$

The coefficients $c_{ir}$ thus introduced are determined by a variational calculation so as to minimize the energy of the system.

The choice of the basis functions $\chi_r$ is rather arbitrary. Different types of functions have been proposed and discussed [31]. We shall not consider this delicate problem. A current choice is to take the functions or *atomic orbitals* associated with electrons in the isolated atoms. For simplicity, we shall take the latter in their real forms.

It is quite obvious that the greater the number of $\chi_r$ functions used, the better the representation (2) of molecular orbitals will be. However, in practice, only a relatively small number of functions $\chi$ can be introduced. The experience of calculations shows that in general a good approximation is obtained if the atomic orbitals used correspond to the energy levels populated by electrons in the atomic ground states [32]:

$1s$ for H; $1s$, $2s$, $2p_x$, $2p_y$, $2p_z$, for first-row atoms, etc.

Once the basis functions have been chosen, explicit determination of the $\varphi$'s is only a matter of computation. The classical method is Roothaan's self-consistent-field (S.C.F.) [33], which modern computers make easily applicable.

## C. Electron Density in a Molecule

Knowledge of the wave function $\psi$ permits calculation of the various quantities which characterize the molecule. In particular the total electron density appears in the form:

$$\varrho^{e} = 2 \sum_i \varphi_i^2 \tag{3}$$

the factor 2 coming from the fact that each orbital $\varphi_i$ is associated with two electrons.

This expression shows that the contribution to the total density of the pair described by the molecular orbital $\varphi_i$ is $2\varphi_i^2$. Knowledge of the total electron density in every point of space permits the calculation of the center of gravity of the electron cloud. As the position of the center of gravity of the positive charges of the nuclei is easily calculated, the electric dipole moment of the molecule can be immediately derived. On the other hand, the charges as postulated by chemists do not follow directly, for it is not possible to identify a discrete set of charges with the continuous distribution given by calculations.

Another difficulty arises from the quantum representation. Strictly speaking, the electronic cloud extends to infinity, so that a reactant attacking the molecule is never situated outside the electron cloud, *i.e.* outside the molecule. However, owing to the rapid decrease of the electron density as the distance from the nuclei increases, it is justified in practice to assume that the electron cloud is contained in a well-defined finite volume **V** having dimensions of the same order as the internuclear distances. For instance, 99% of the electron cloud of the $H_2$ molecule is located inside a revolution ellipsoid with axis 4.5 and radius 1.6 Å. The picture of the molecule as a finite size object, as viewed in classical chemistry, is thus justified even though it is only a very good approximation. The fact that the electron cloud extends to infinity does not produce mathematical difficulties concerning the definition of the electrostatic potential created at any point of space. In fact, the potential created in its center by a charged sphere whose charge density is everywhere finite ($|\varrho| \leqslant |\varrho_m|$) and of the same sign, tends to zero with the radius $a$ of the sphere:

$$V = \int_0^a \frac{|\varrho|}{r} 4\pi r^2 \, dr \leqslant |\varrho_m| \, 2\pi a^2 .$$

## III. Equivalent Multipoles

As has been mentioned, the electrostatic potential created by the molecule plays an important role in the ionic reactions as well as in molecular interaction phenomena. As the problem has been presented and discussed in detail in an article recently published in this series [34], we shall not discuss it here.

It is quite obvious that the best representation of this potential consists in constructing contour lines maps (Figs. 1, 2). Such pictures are often very effective, but have the limitation that several sections are necessary to represent the potential in space, and they do not lend themselves easily to further calculations. These diagrams, sometimes not easy to interpret, will be discarded by the chemist in favor of a more tangible and familiar description, based on point charges and electric multipoles, especially dipoles. The problem is then to reproduce as well as possible the electrostatic field created by the molecules by means of these point charges and these dipoles.

### A. General Expression of the Electrostatic Potential Created by a Molecule

The potential created by a molecule is completely defined by the charge density values in every point of space. This density may be written:

$$\varrho(M) = -\varrho^{e}(M) + \sum_K N_K \, \delta(\mathbf{M} - \mathbf{K}) \tag{4}$$

where $\varrho^e(M)$ is the electronic density at point M, defined by the vector **M**, $N_K$ is the positive charge of the nucleus whose position vector **K**, and $\delta$ is the Dirac distribution.

In the following, we shall always assume that the nuclei are fixed. It is well known that the molecule is actually in a perpetual state of vibration, the nuclei oscillating around equilibrium positions corresponding to the minima of energy. We shall suppose that the nuclei are fixed in these positions and we shall define the coordinate system $O_{xyz}$ with reference to them.

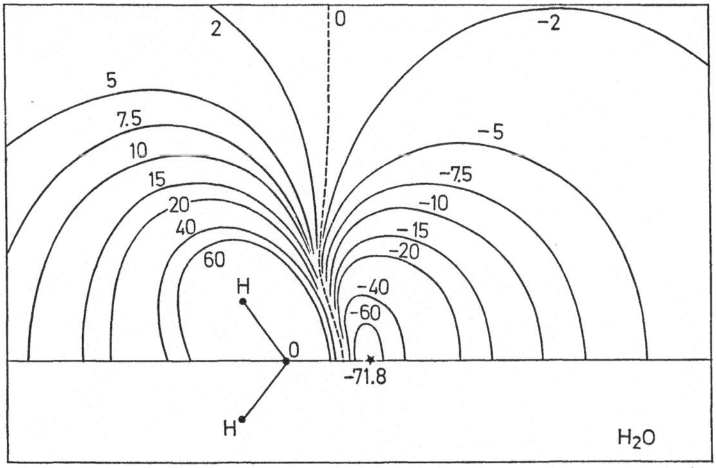


Fig. 1. Electrostatic potential-energy map for $H_2O$ in the molecular plane [34]. Values are expressed in kcal-mole

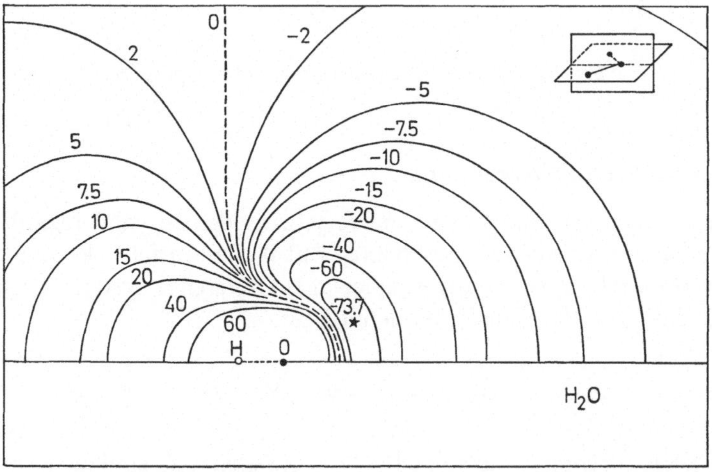


Fig. 2. Potential-energy map for $H_2O$ in the symmetry plane perpendicular to the molecular one [34]

The potential created at a point P situated at a distance $R$ from the origin (OP$=R$), in the direction specified by cosines $\alpha$, $\beta$, $\gamma$, may be expanded to the powers of $R^{-1}$ according to the classical expression:

$$V_P(R;\alpha,\beta,\gamma) = \frac{1}{R}\int \varrho dv + \frac{1}{R^2}\left[\alpha\int \varrho x dv + \beta\int \varrho y dv + \gamma\int \varrho z dv\right] \quad (5)$$
$$+ \frac{1}{2R^3}\left[\alpha^2\int \varrho\left(2x^2 - y^2 - z^2\right) dv + \ldots + 6\alpha\beta\int \varrho xy dv + \ldots\right] + \ldots$$

(the integrations being taken over the whole space).

The term in $R^{-(n+1)}$ of Eq. (5) corresponds to the potential created by a multipole of order $2^n$.

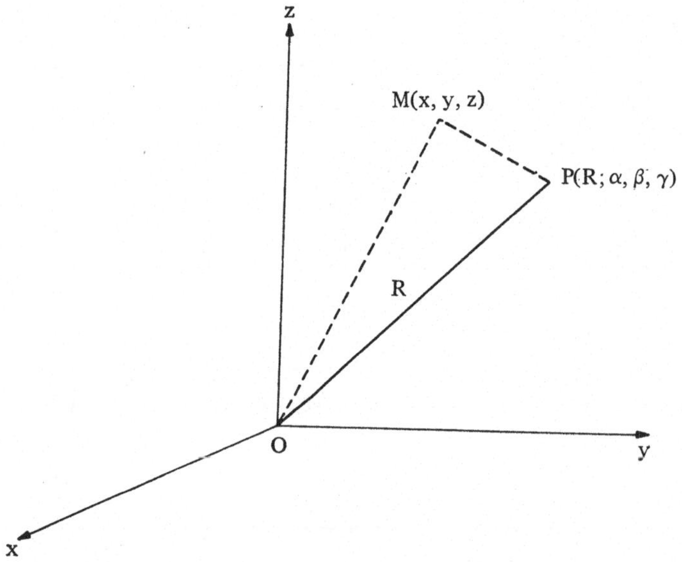


Fig. 3. Notation for the expansion of the potential

## B. Study of Multipole Expressions

The expansion Eq. (5), taken in its entirety, is evidently valid regardless of the origin or the orientation of the chosen axis. But it is essential to keep in mind that the various integrals depend on the choice of the reference system. To see this we perform for instance the translation defined by:

$$x \to x' = x - X, \; y \to y' = y - Y, \; z \to z' = z - Z\,.$$

Then, $\int \varrho x dv$ becomes $\int \varrho x' dv = \int \varrho x dv - X \int \varrho dv$, and similar formulas hold for $y$ and $z$. Thus, if the total charge $\int \varrho dv$ does not vanish, the dipole components $\int \varrho x dv, \ldots$, are not invariant upon translation.

The same kind of conclusion holds for the quadrupole terms:

$$\int \varrho xy dv \rightarrow \int \varrho x'y' dv = \int \varrho xy dv - X \int \varrho y dv - Y \int \varrho x dv + XY \int \varrho dv$$

$$\int \varrho (2x^2 - y^2 - z^2) dv \rightarrow \int \varrho (2x^2 - y^2 - z^2) dv - 4X \int \varrho x dv + 2Y \int \varrho y dv + 2Z \int \varrho z dv + (2X^2 - Y^2 - Z^2) \int \varrho dv$$

these quantities and their analogs are not invariant unless the total charge and the quadrupole components vanish.

In general, *only the components of the first non-vanishing multipole are invariant.* This indicates that it is always necessary to specify the choice of the reference system used in the calculations.

This lack of invariance may seem at first to represent a serious difficulty if the exact expression of the potential must be replaced by a limited expansion. Actually, it can be taken advantage of by choosing the coordinate system so as to simplify the expansion as far as possible. We consider a few specific cases:

1. Suppose the total charge is not zero, $\int \varrho dv \neq 0$, as in the case of an ion. One can choose as the origin the center of gravity of the charge distribution. In any reference system with origin in that point, the dipole term vanishes, hence:

$$\int \varrho x dv = \int \varrho y dv = \int \varrho z dv = 0 .$$

Moreover, by convenient choice of the $x$, $y$, $z$ directions, the rectangular contributions $\alpha\beta$, $\beta\gamma$, $\gamma\alpha$ appearing in the quadrupole term may be made to vanish. Hence the simplest expression that can be obtained will be:

$$V_P(R; \alpha, \beta, \gamma) = \frac{1}{R} \int \varrho dv + \frac{1}{2R^3} \left( A\alpha^2 + B\beta^2 + C\gamma^2 \right) + \ldots \quad (6)$$

with

$$A = \int \varrho (2x^2 - y^2 - z^2) dv, \ \ldots \text{ and } A + B + C = 0$$

When the charge distribution is cylindrically symmetric around the $z$-axis (linear ions), $A = B = \int \varrho (x^2 - z^2) dv$, and

$$V_P(R; \alpha, \beta, \gamma) = \frac{1}{R} \int \varrho dv + \frac{3\gamma^2 - 1}{2R^3} \int \varrho \left( z^2 - x^2 \right) dv + \ldots \quad (7)$$

Any choice of the origin other than the center of gravity of the charges will give rise to a dipole term.

2. Suppose the total charge is zero: $\int \varrho dv = 0$. This is the case in a neutral molecule.

The center of gravity of the whole system can no longer be defined. But a center of gravity $G^+$ of the positive charges and a center of gravity $G^-$ of the negative charges can be separately defined. Suppose these two points do not coincide. Choose as the $z$-axis the straight line joining $G^+$ and $G^-$; then:

$$\int \varrho x dv = \int \varrho y dv = 0$$

for any origin lying on this straight line and for any choice of the $x$ and $y$ axes around it.

It is also easy to verify that in a translation along the $z$ axis — the direction of $x$ and $y$ remaining the same — the quantities:

$$\int \varrho z dv, \int \varrho xy dv, \int \varrho xz dv, \int \varrho yz dv, \int \varrho x^2 dv, \int \varrho y^2 dv$$

are invariant. The potential thus takes the following form:

$$V_P(R;\alpha,\beta,\gamma) = \frac{\gamma}{R^2}\int \varrho z dv + \frac{1}{2R^3}\Big(A\alpha^2 + B\beta^2 + C\gamma^2 + 2D\alpha\beta + 2E\beta\gamma + 2F\gamma\alpha\Big) + \ldots \tag{8}$$

where $D$, $E$, $F$ are independent of the origin, whereas A, B, C are functions of it. The axes of the quadric $Ax^2 + By^2 + \ldots + 2Fzx = 1$ which corresponds to the quadrupole term are not oriented in general in any special way with respect to the $G^+G^-$ line. Therefore in general it will not be possible to give a simple expression of $V_P$.

An interesting particular case is that of a charge distribution belonging to the $C_{2v}$ group (*e.g.* the $H_2O$ molecule). Such a system has two perpendicular symmetry planes. The $G^+G^-$ straight line necessarily coincides with the intersection of these two planes. For symmetry reasons it is best to take these planes as the $x=0$ and the $y=0$ planes. The expression of the potential then reduces to:

$$V_P(R;\alpha,\beta,\gamma) = \frac{\gamma}{R^2}\int \int \varrho z dv + \frac{1}{2R^3}\Big(A\alpha^2 + B\beta^2 + C\gamma^2\Big) + \ldots$$

In a translation $z \to z' = z - Z$, $\int \varrho x^2 dv$ and $\int \varrho y^2 dv$ remain constant, $\int \varrho z^2 dv$ becomes $\int \varrho z^2 dv - 2Z\int \varrho z dv$. The non-invariance of the coefficients $A$, $B$, $C$, allows us to impose an additional condition.

For instance, we can try to have $(3\gamma^2 - 1)$ as a factor. This demands $A = B$, *i.e.* $\int \varrho x^2 dv = \int \varrho y^2 dv$. Now this is not realized in general, except if the system is one of revolution around the $z$-axis. In this case the potential expansion is:

$$V_P(R;\alpha,\beta,\gamma) = \frac{\gamma}{R^2}\int \varrho z dv + \frac{3\gamma^2 - 1}{2R^3}\int \varrho\Big(z^2 - x^2\Big)dv + \ldots \tag{9}$$

In a translation $z \to z' = z - Z$, $\int \varrho(z^2 - x^2)dv$ becomes $\int \varrho(z^2 - x^2)dv - 2Z\int \varrho z dv$, and thus the origin can be chosen so as to make the quadrupole term vanish. But it is essential to keep in mind that in this case the center of the dipole (the origin) will not coincide in general with the center of the $G^+G^-$ segment, as one might have expected. It is easy to verify this point directly on the simple example of two $-1$ and one $+2$ charge, arranged as in Fig. 4:

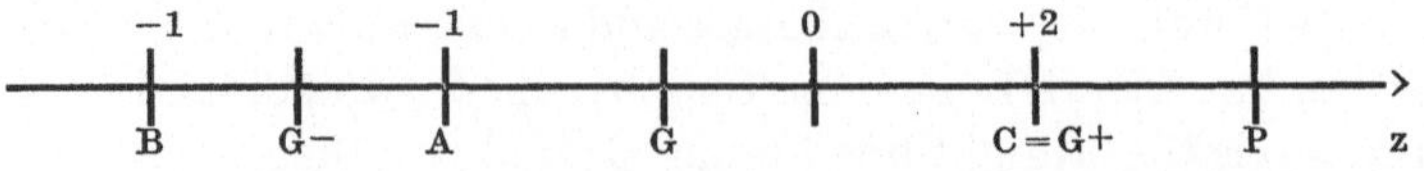


$|AC| = a$, $|BC| = b$, $|CG| = (a+b)/4$, $|OC| = Z$, $|OP| = R$.

Fig. 4. Example of charge distribution. Notation

Taking the center of the $G^+G^-$ as the origin, in a point P of the axis, we have:

$$V_P^G = \frac{a+b}{R^2} - \frac{(a-b)^2}{2R^3} + (R^{-4})$$

whereas if the origin is such that $Z = (a^2 + b^2)/2(a+b)$, we have simply:

$$V_P^O = \frac{a+b}{R^2} + (R^{-4})$$

In the case when the charge distribution has no cylindrical symmetry, and the $A = B$ equality cannot be satisfied, one can have $B = C$ by choosing the origin so that:

$$\int \varrho y^2 dv = \int \varrho (z - Z)^2 dv, \text{ i.e. : } 2Z \int \varrho z dv = \int \varrho (z^2 - y^2) dv$$

The potential is then:

$$V_P(R; \alpha, \beta, \gamma) = \frac{\gamma}{R^2} \int \varrho z dv + \frac{3\alpha^2 - 1}{2R^3} \int \varrho \left(x^2 - z^2\right) dv + \dots \tag{10a}$$

The formula introduces a term which corresponds to a quadrupole having the same center as the dipole and lying along the $x$-axis.

Choosing the origin so as to make $A = C$, one obtains in the same way:

$$V_P(R; \alpha, \beta, \gamma) = \frac{\gamma}{R^2} \int \varrho z dv + \frac{3\beta^2 - 1}{2R^3} \int \varrho \left(y^2 - z^2\right) dv + \dots \tag{10b}$$

Any other choice of the origin would give less simple expressions. In particular, to take the center of $G^+G^-$ as the origin would not be very convenient.

3. With high symmetry systems, such as $MX_6$, the highest number of vanishing terms is obtained by choosing as the reference axes the symmetry axes of the system (here the MX straight lines). The potential expression reduces to:

$$\begin{aligned} V_P(R; \alpha, \beta, \gamma) &= \frac{1}{R} \int \varrho dv \\ &+ \frac{7}{4R^5} \left(\alpha^4 + \beta^4 + \gamma^4 - 3\alpha^2\beta^2 - 3\beta^2\gamma^2 - 3\gamma^2\alpha^2\right) \\ &\int \varrho \left(x^4 - 3x^2y^2\right) dv + \dots \end{aligned} \tag{11}$$

In conclusion, we see that a proper choice of the reference system may introduce considerable simplification of the potential created by a molecule.

### C. Representation of the Potential by Means of a Finite Number of Multipoles

The expansion of the electrostatic potential in terms of multipoles has the advantage of providing an analytic representation of equipotential surfaces, but, of course, it will be of practical interest only if the expansion converges rapidly enough for a small number of terms to be important.

Direct comparison of the exact expression and a limited multipole expansion of the potential suggests that hexadecapole terms (in $R^{-5}$) give a negligible contribution with respect to lower order terms [35].

If the distance is sufficiently large, the octupole terms can also be neglected without any serious error. At any rate, expansion limited to the quadrupole term always represents the shape of the potential in a satisfactory qualitative manner. In practice, inclusion of quadrupole terms is sufficient for problems related to chemical reactivity.

It follows that a reasonable choice of the reference system will make it possible to describe, from the electrostatic point of view:

- an ion by means of a point charge and a quadrupole;
- a neutral molecule of any kind by means of a dipole, possibly vanishing, and of a quadrupole with the same center;
- a linear neutral molecule by means of a dipole; etc.

As concerns interactions between molecules, this type of approximation will be sufficient as long as gases are considered. It may not be so for the condensed states, such as liquids or molecular crystals; in these cases it will be necessary to take into account higher order terms of the multipole expansion.

## IV. Definition of Charges from Moments

### A. The Principle

The knowledge of the field created by a molecule in terms of multipoles makes possible the discussion of the problem of electrical charges and their meaning. In fact, by definition, the charges must reproduce the electrostatic field created by the molecule. The problem thus consists in determining the values $Q_i$ of the various point charges and the positions $(x_i, y_i, z_i)$ where they must be placed so as to obtain the same multipole contributions. We shall call these charges, positive or negative, *net charges*.

A preliminary remark is in order. In a molecule, charges are of two types: the positive point charges of the nuclei and the diffuse charges of the electrons. However, nuclei other than the proton are surrounded by internal spherical shells; therefore, according to Gauss's theorem, one can only consider the field created by each spherical core (nucleus plus internal electron shells) as the field of a positive point charge. Clearly the total field resulting from the positive charges of the H nuclei and of the cores, and from the negative charges due to valence electrons is involved in interactions with an external point charge. The field of the H nuclei and of the cores is the same in a real molecule as in the dicrete-charge description; therefore, it will be enough to consider the potential created by the binding electrons in order to identify the total potential in the two cases. We shall note the electronic charges $q$. These charges are negative. ($q$ will be positive in electronic unity). If the positions of the electronic charges ($q$) coincide with those of the nuclei (positive charges $N$):

$$Q = N - q\,.$$

## B. Determination of the Values and Positions of the Charges

The values $q_i$ of the various charges $i$ and their positions ($x_i$, $y_i$, $z_i$) must satisfy the following relationships:

a) $\sum q_i = n$, with $n$ the total number of the valence electrons;

b) $\sum q_i x_i = \int \varrho x dv$ and the corresponding equalities for y and z;

c) $$\sum q_i (2x_i^2 - y_i^2 - z_i^2) = \int \varrho (2x^2 - y^2 - z^2)\, dv \qquad (12)$$
$$\sum q_i (2y_i^2 - x_i^2 - z_i^2) = \int \varrho (2y^2 - x^2 - z^2)\, dv$$
$$\sum q_i x_i y_i = \int \varrho xy dv,\; \sum q_i y_i z_i = \int \varrho yz dv,\; \sum q_i z_i x_i = \int \varrho zx dv$$

etc. for the higher orders ($\varrho$ is here the electron density due to the valence electrons).

We have thus:

1 equation for the unipole term;

3 equations for the dipole term;

5 equations for the quadrupole term;

In general, we have $(2n+1)$ relations for the term of order $2^n$. Now for the $i$—th charge we have four unknowns, its value $q_i$ and its three coordinates. If the number of charges to be considered is fixed *a priori*, it will always be possible to obtain a sufficient number of equations to determine the unknowns uniquely by introducing a sufficient number of multipole terms. But evidently nothing will guarantee that the values of the charges and the corresponding coordinates thus obtained will verify the equations for the multipole terms of higher order.

In any case the problem of molecular charges includes many special constraints, for in the mind of the chemist the charges are viewed as located near if not on the nuclei. Therefore, their number is limited and fixed in advance as being the same as the number of nuclei in the molecule. Of course, this restriction will have a great impact on the quality of the description, for it will be possible to equate only a small number of multipole terms. The increase in the number of point charges introduced will improve the description. For instance, for the molecule $H_2O$, with 13 charges, one obtains a potential which practically coincides with the exact potential [36].

However, such a model does not interest the chemist. Therefore in the following, we shall always assume that the number of charges is the same as that of the nuclei.

Let us examine first the case of a diatomic molecule AB. Let $q_A$ and $q_B$ be the charges. For symmetry reasons we shall place them on the nuclear axis. Let $z_A$ and $z_B$ be their non-vanishing coordinates. Then:

$$\begin{cases} q_A + q_B = n \\ q_A z_A + q_B z_B = \int \varrho z dv = \lambda_1 \\ q_A z_A^2 + q_B z_B^2 = 2 \int \varrho (z^2 - x^2)\, dv = \lambda_2 \\ \cdots\cdots\cdots\cdots \end{cases}$$

The first two equations give:

$$q_A = \frac{\lambda_1 - n z_B}{z_A - z_B} \text{ and } q_B = \frac{-\lambda_1 + n z_A}{z_A - z_B}. \tag{13}$$

These values are invariant in a translation $z \to z' = z - Z$.

In accordance with the traditional chemical view, the negative point charges are located on the corresponding nuclei A and B, the $q$ values are then perfectly determined. But, in general, the third relationship involving the quadrupole term will not be satisfied. The dipole and the quadrupole term can only be reproduced completely if the electronic charges are located at points other than the nuclear positions. Furthermore, for determining $z_A$ and $z_B$ it will be necessary to have recourse to the equation for the octupolar term, which will thus also be reproduced.

When the molecule is formed by a sufficient number of nuclei, the dipole, quadrupole and even octupole terms can be made to coincide even assuming that the charges are located on the nuclei. For instance, with three nuclei on the same straight line, the following three equations are sufficient for determining uniquely the charges $q_A$, $q_B$, $q_C$ centered on the nuclei A, B, C with abscissas $z_A$, $z_B$, $z_C$:

$$\begin{cases} q_A + q_B + q_C = n \\ q_A z_A + q_B z_B + q_C z_C = \lambda_1 \\ q_A z_A^2 + q_B z_B^2 + q_C z_C^2 = \lambda_2 . \end{cases}$$

The quadrupole moment is then reproduced.

In the case of systems with a high degree of symmetry, the identification can be made with higher order multipoles. For instance, in molecules of the $MX_6$ type, where for symmetry reasons a charge $q'$ will be located at the central nucleus M and six equal charges $q$ will be located on the $MX_6$ lines at equal distances from M, we shall have:

$$V_P = \frac{q' + 6q}{R} + \frac{7qa^4}{2R^5} (\alpha^4 + \beta^4 + \gamma^4 - 3\alpha^2\beta^2 - 3\beta^2\gamma^2 - 3\gamma^2\alpha^2) + \ldots \tag{14}$$

Hence, by identification with the potential created by the continuous real distribution [Eq. (11)]:

$$2qa^4 = \int \varrho (x^4 - 3x^2y^2)\, dv\,.$$

If the charges $q$ are required to be located at the X nuclei, the $q$ value is obtained. Only by using the subsequent multipole term, which gives the value of the $qa^6$ product, would it be possible to determine both $q$ and $a$. However, in view of the order already reached in the expansion, such a calculation is not especially useful, and the charges $q$ can be assumed to be located at the X nuclei.

These few examples clearly show that to require *a priori* that net charges in a molecule should be located at the nuclei is merely an approximation not acceptable in the case of diatomic molecules, but acceptable for the other molecules, so that the traditional chemical view appears to be justified.

## V. Determination of Charges from the Cohesion Energy of an Ionic Crystal

A very special but interesting case is given by symmetry ions of the $AB_n^{m-}$ type in an ionic crystal. The cohesion of the structure is explained in the classical theory [37] by equilibrium between the electrostatic forces, whose resultant is attractive, and the short-range repulsive forces acting between ions.

In the case when all the ions are mononuclear, as in NaCl, the electron distribution around each nucleus is isotropic, so that one can assume that the electrostatic energy results from the interaction of point charges centered on the nuclei. This is Madelung's classical calculation [38].

If the ions are not single nuclei as we have seen, in the calculation of the electrostatic potential created by them, the continuous charge distributions can be replaced by point charges located on their nuclei. In the case of symmetric ions of the type $AB_n^{m-}$, such as $NO_3^-$, $CO_3^{2-}$, $SO_4^{2-}$, ..., the net charges carried by the B nuclei can be expressed in terms of the charge carried by the central atom:

$$Q_B = -\frac{m + Q_A}{n}\,.$$

Knowledge of the geometry of the ion and of the geometry of the crystal thus makes possible calculation of the electrostatic energy, and therefore the cohesion energy of the crystal as a function of the charge $Q_A$. Comparison with experiment then gives $Q_A$ und $Q_B$. For instance, in lithium nitrate one finds for the nitrate ion [39]:

$$Q_N = 1.16 \quad \text{and} \quad Q_0 = -0.72\,.$$

## VI. Direct Definition of Electronic Charges from the Wave Function

Instead of trying to derive the charges from the various equivalent multipoles, one can try directly to use the wave functions describing the electrons in the molecule. The problem is neither simple nor easy.

### A. Basic Convention

We have given above the expression of the electron density in terms of molecular orbitals [Eq. (3) of Section III-C.]. The center of gravity of the electron cloud, $\mathbf{G}^-$ will be given by the expression:

$$n\mathbf{G}^- = \sum 2\int \varphi_i^2 \, \mathbf{M} \, dv = 2 \sum_i \sum_r \sum_s c_{ir} \, c_{is} \int \chi_r \, \mathbf{M} \, \chi_s \, dv \tag{15}$$

where $n$ is the total number of electrons and $\mathbf{M}$ is an arbitrary point of space.

By definition, we call *population* of an orbital $\chi_r$ any quantity $q_r$ such that

$$n\mathbf{G}^- = \sum_r q_r \, \mathbf{I}_r \tag{16}$$

where $\mathbf{I}_r$ is the position vector of the center of gravity $\int \chi_r \, \mathbf{M} \, \chi_s \, dv$ of the cloud associated with the function $\chi_r$. When the functions $\chi_r$ are atomic orbitals, the $\mathbf{I}_r$s coincide with the position vectors of the nuclei carrying the corresponding orbitals. But in certain problems it is sometimes interesting to use not the atomic orbitals as such, but linear combinations of orbitals carried by the same nucleus (hybrid orbitals [40]). The center of gravity lies then in a slightly excentric position with respect to the nucleus to which the hybrid orbitals belongs [41].

Even functions whose centers do not coincide with the nuclei (floating orbitals [42]) can be used. However, to remain within the chemist's picture, *i.e.* to obtain charges located on the nuclei or near them, it is more convenient to use either atomic orbitals centered on the nuclei, or hybrids constructed therewith. This we shall assume in the following.

The populations $q_r$ which we shall define will thus give the charges carried by the various orbitals $\chi_r$ and will be located at the centers of gravity of the corresponding distributions $\chi_r^2$. Thus stated, the problem of defining the charges is undetermined. In fact, an infinite number of different point-charge systems can have the same center $\mathbf{G}^-$ as the overall cloud. In order to reduce the degree of arbitrariness, one can try to express the various integrals $\int \chi_r \, \mathbf{M} \, \chi_s \, dv$ involving two centers as sums of one-center terms. However the identification does not call into play multipolar terms of order higher than two; therefore, the charges thus obtained will not necessarily have the properties implied in the chemist's model and will be subject to criticism. This is related to the fact that, in the chemist's picture, charges do not belong to the category of those quantities which quantum theory calls "observable" [43], *i.e.* defined more or less directly in terms of a physical measurement.

### B. Examples of Approximations

The oldest approximation [44] is perhaps the one introduced for $\pi$ orbitals, *i.e.* orbitals with parallel axes, carried by nuclei located in the same plane. This case is encountered with planar conjugated molecules. The center of gravity of the $\chi_r \chi_s$ distribution is assumed to lie at the center of the segment joining the nuclei to which the orbitals $\chi_r$ and $\chi_s$ belong. As these nuclei are the centers of gravity of the distributions $\chi_r^2$ and $\chi_s^2$, one has:

$$\int \chi_r \mathbf{M} \chi_s \, dv = \frac{1}{2} (\mathbf{I}_r + \mathbf{I}_s) S_{rs} \tag{17}$$

$S_{rs}$ being the overlap integral $\int \chi_r \chi_s \, dv$.

It follows that

$$n\mathbf{G}^- = \sum_i \sum_r c_{ir} c_{is} (\mathbf{I}_r + \mathbf{I}_s) S_{rs}/2 = 2 \sum_i \sum_r \sum_s c_{ir} c_{is} S_{rs} \mathbf{I}_r$$

whence

$$q_r = 2 \sum_i (c_{ir}^2 + \sum_{s \neq r} c_{ir} c_{is} S_{rs}) \tag{18}$$

A more sophisticated treatment consists in a better description of the position of the center of gravity of $\chi_r \chi_s$ by writing [45]:

$$\int \chi_r \mathbf{M} \chi_s \, dv = \frac{1}{2} (\lambda_{rs} \mathbf{I}_r + \mu_{rs} \mathbf{I}_s) S_{rs} \tag{19}$$

with

$$\lambda_{rs} + \mu_{rs} = 2 \, .$$

It follows that:

$$q_r = 2 \sum_i (c_{ir}{}^2 + \sum_{s \neq r} c_{ir} c_{is} S_{rs} \lambda_{rs}) \tag{20}$$

However, in the case of $\pi$-type orbitals, the parameters $\lambda$ and $\mu$ are close to unity. They are rigorously equal if the orbitals have the same nature, as the $2p\pi$ orbitals of carbon. Their values are 1,22 and 0,78, respectively, for the $2p\pi$ orbitals of carbon and oxygen in the carbonyl group ($d = 1{,}21$ Å). Therefore, Eqs. (18) and (20) give slightly different results: $q_0 = 1{,}13$ from Eq. (18) vs. $q_0 = 1{,}17$ from Eq. (20) (Calculation from the results Ref. [46]).

The charges thus defined are centered on the nuclei. However the relationships of Eq. (17) and Eq. (19) are but approximate ones. To improve them, one might for instance replace the distribution $\chi_r \chi_s$ by an expression of the type $(a\chi_A'^2 + b\chi_B'^2) S_{rs}$ where $\chi_A'$ and $\chi_B'$ represent appropriate orbitals centered in points A and B different from the nuclei carrying the $r$-th and the $s$-th orbital. The position of these points as well as the explicit expression of the orbitals $\chi_A'$ and $\chi_B'$ is determined so as to obtain coincidence of quadrupole and octupole terms in addition to the center of gravity [47]. Other decompositions are also possible [48]. The decomposition of the orbital product $\chi_r \chi_s$ is thus improved, but this has the inconvenience of introducing orbitals which are excentric with respect to their nuclei and are thus in disagreement with the chemist's picture. Therefore, these approximations are not used for defining the charges.

Although originally introduced for $\pi$ orbitals, Eq. (20) also applies to a system formed solely by $s$ orbitals, or to a set of orbitals having a common axis of revolution, as in a linear molecule. In the case of $s$ or $p$ orbitals, the charges are still located at the nuclei. In the case of hybrids orbitals, they are excentric with respect to the nuclei, but they are situated on the line joining them.

A limiting case which fits in the general frame of Eq. (19) is when the $\chi_r \chi_s$ distribution is highly unsymmetric, as happens in a strongly polar bond. Then $\lambda_{rs}$ is very close to 2 and $\mu_{rs}$ is very close to 0, or *vice versa*.

One has, for instance:

$$\int \chi_r \mathbf{M} \chi_s \, dv = \mathbf{I}_r S_{rs}$$

Such an approximation leads, in the instance of a wave function only consisting of the orbitals $\chi_r$ and $\chi_s$, to the charges [49]:

$$\begin{cases} q_r = 2c_r^2 + 4c_r c_s S_{rs} \\ q_s = 2c_s^2 \,. \end{cases}$$

Eqs. (18) and (20) have been established in a very special case. They hold only if the centers of gravity of all the distributions $\chi_r \chi_s$ are situated on the straight lines joining the centers of gravity of the distributions $\chi_r^2$ and $\chi_s^2$. Usually this favorable circumstance does not take place, and it is difficult to obtain simple equations. Nevertheless, it has become customary [50] to define in all cases the charges carried by the various orbitals by means of Eq. (18). The latter is approximately valid, as has been shown, for certain types of molecules and orbitals, but leads in the other cases to values which are meaningless and therefore cannot but generate confusion in the mind of the chemist. For instance, in benzene, the SCF method gives a net charge $-0{,}20$ on the carbon atom [51], whereas one would have expected a value close to zero. This formula can even give *negative* electron populations, as is the case in $Cu\,Cl_4^{2-}$ [52].

Other definitions have been proposed to avoid such difficulties [53]. Their significance is less easy to grasp, and we shall not dicusss them here.

### C. Effect of Orthogonalization of Basis Orbitals

Instead of atomic orbitals or hybrids, which have the disadvantage of not being orthogonal to one another, use is often made of linear combinations there of, made orthogonal to one another by the procedure of Landshoff [54], generalized by Löwdin [55], which can be symbolically written as:

$$(\chi') = (1 + \mathbf{S})^{-1/2} (\chi) \tag{21a}$$

$\mathbf{S}$ being the matrix whose general element is: $\mathbf{S}_{pq} = S_{pq} - \delta_{pq}$.

To the first order in $S$, Eq. (21a) becomes:

$$\chi_r' = \chi_r - \frac{1}{2} \sum_{k \neq r} S_{kr} \chi_k \,. \tag{21b}$$

With this orthogonal basis, Eqs. (18) and (20), when applicable, take the simpler form [55]:

$$q'_r = 2 \sum_i (c'_{ir})^2 \tag{22}$$

$c'_{ir}$ being the new coefficients of the molecular functions $\varphi'_i$ built on the orthogonal $\chi'$ basis. The charge thus obtained is evidently strictly positive, so that the difficulty encountered with Eq. (19) in connection with the sign of $q$ does not appear. This may seem contradictory, for the charges $q'_r$ are derived from the coefficients $c'_{ir}$ which are related to the initial coefficients $c_{ir}$. Actually, $q'_r$ represents the charge carried by the distribution $\chi'^2_r$. Now, according to Eq. (21a) and Eq. (21b), the corresponding cloud is delocalized on the whole set of the atomic orbitals $\chi_r$. Therefore, the center of gravity of $(\chi'_r)^2$ does not coincide with that of $\chi^2_r$, and the charges $q'_r$ and $q_r$ thus defined, although they give the same overall description, do not have the same meaning. A favorable case is when the charges come from $\pi$-type orbitals. The charges $q$ are obtained from the charges $q'$ by a mathematical process which consists in inverting the orthogonalization procedure. To the second order in $S$, one obtains [56]:

$$q_r = q'_r + \frac{1}{4} \sum_{s \neq r} (q'_r - q'_s)\, S_{rs}^2 \,. \tag{23}$$

The relationship shows that the charges $q$ and $q'$ are not much different, for $S^2$ never exceeds 0.1 in this case, so that at a first approximation one can give the charges the $q'$ values and consider them as being located at the corresponding nuclei. For orbitals of a general type, also neglecting $S^2$, one obtains: $q_r = q'_r$ [57].

Special mention must be made of semiempirical methods of the Hückel type: the Hückel method proper for $\pi$-systems [30] and the so-called Extended Hückel method [58] for molecules in general. In these methods the atomic orbitals are assumed to be orthogonal, and the charges are given by Eq. (22) [59]. In these case of $\pi$-systems, the equivalence of these methods with the more rigorous SCF method can be proven [60], by the precise use of the Löwdin orthogonalization procedure. In the other cases, the value of these methods is more dubious.

### D. A Defect Characteristic of the Above Definitions

Regardless of the special definition adopted, one difficulty makes the values obtained for the charges subject to criticism [43]. Even when the center of gravity of the electron cloud associated with a given atomic orbital coincides with the corresponding nucleus, the charge is often concentrated in points rather far from the latter. For instance, in the case of a $2p$ orbital with effective charge $Z$, the maximum is located at $2a_0/Z$ ($a_0$ being the Bohr radius, 0,5293 Å). This phenomenon is even more pronounced with hybrid orbitals presenting a density concentrated in a given direction. It follows that the density maximum of a given orbital may lie close to another nucleus. Therefore, a calculation which assigns the whole charge carried by a given orbital to the corresponding nucleus will tell us nothing about the fraction of electrons actually present in the neighborhood of the nucleus. The theoretical values will have little in common with the charges required by chemists.

One solution consists in using the electron density maps to determine numerically the fraction of electrons contained in a given region around the nucleus. This procedure is difficult to realize because of the difficulty of finding a general definition of the region to be assigned to each nucleus. The difficulty could be overcome by referring to some physical quantity whose observed values are concentrated around the different nuclei, except for linear [61] or monocyclic molecules [62].

### E. Evaluation of Charges from the NMR Chemical Shift

The study of chemical shifts in NMR provides an example of the sort of physical quantities to which one may refer. As is well known, the total magnetic field to which a nucleus of a molecule is subject is the sum of the external field $\mathscr{H}_{\text{ext}}$ and of the induced field created by the electrons surrounding the nucleus under study. The latter field is proportional to the external field and of opposite sign. The effective field thus acting on the nucleus is:

$$\mathscr{H}_{\text{eff.}} = \mathscr{H}_{\text{ext}}\,(1 - \sigma) \tag{24}$$

$\sigma$ being the magnetic shielding constant. This constant depends essentially on the electron distribution in the immediate neighborhood of the given nucleus. The resonance condition for a nucleus of a given species is always obtained for the same value of the effective field; therefore, the external field required to obtain resonance must depend on $\sigma$, *i.e.* on the environment of the nucleus. This is the essence of the *chemical shift effect*.

In a series of similar compounds, the $\sigma$ constant may be related to the electric charge present around the nucleus, and indirectly to that of the neighboring atoms. For instance, for a proton carried by a carbon atom participating in a conjugated system, the chemical shift will depend on the net $\pi$ charge of the carbon atom. If the carbon atom is positively charged, its electronegativity is higher [63], and it attracts the electrons more than if it were neutral. Experiment provides the values of the chemical shifts for three molecules where the net charges $Q$ are known beforehand for symmetry reasons: the cyclopentadienyl anion $C_5H_5^-$ ($Q = -1/5$), benzene $C_6H_6$ ($Q = 0$), and the tropylium ion $C_7H_7^+$ ($Q = 1/7$). By interpolation one obtains an exponential law $\sigma = \sigma(Q)$ [64]. With this law, measurement of $\sigma$ in a hydrocarbon allows determination of the charge $Q$. Unfortunately, this procedure does not apply to all cases, and moreover the phenomenon is more complicated; the value of the chemical shift depends on other structural effects (ring currents, hybridization changes with number of carbon atoms in the ring, etc.), and also on interactions with the solvent. Nevertheless, the method has been successfully applied to azulene $C_{10}H_8$, whose net charges are important and where the theoretical predictions have been confirmed [64–65].

On the other hand, several attempts have been made to connect chemical shifts with charge distributions given by semiempirical methods whose parameters have been adjusted to other molecular properties. A linear condition of the type $\sigma = -AQ_C - BQ_H + C$ has been found to be quite satisfactory for the chemical shift of the proton in aminoacids [66]. For $^{13}C$ the results of similar attempts have not been equally encouraging [67], the reason possibly being the charge definition adopted. With an appropriate definition of the type Eq. (20), a good linear correlation between the chemical shift and the net charge of the $^{13}C$ nucleus is obtained [68]. Moreover, as could be expected, the shielding constant obtained is linearly correlated with Taft's polar constants [68].

### F. Relation to ESCA

As will been seen later (VII-Å), the internal shells of atoms are practically not involved in bond formation. Therefore, it is legitimate to speak of the works of extraction of the internal electrons of the various atoms forming a molecule: *e.g.*, the ls electrons of the carbon atom. This energy depends on the quantity of electricity situated in the neighborhood of the atom under study. The comparison between the energy of extraction of an internal electron of an atom in a molecule and in the isolated atom can thus give us information concerning the electronic charge carried by the atom under study [63–70]. The procedure has also been applied to crystals [71].

A linear relationship has also been suggested between the ionization energy of an alkane and the maximum sum of the net charges of two adjacents atoms [70].

### G. Importance of the Quality of the Wave Functions

It is quite clear that the problem of finding a proper definition of atomic charges in a molecule goes together with the problem of exactly determining wave functions. The latter problem has already been mentioned in connection with the calculation of the electrostatic potential created by a molecule.

We shall not repeat these remarks in particular concerning the comparative merits of the various methods.

We should, however, like to call attention to the fact that the various methods may lead to significant differences in the calculated atomic charges. The example of the charges of the alkanes is typical [72]. However, it seems that use of a definition of charges based on Eq. (20), with factors $\lambda$ adapted to the basis, will sensibly lower the discrepancies [68–48].

The case of the $\pi$-methods is more clear cut, for, as has been seen, the definition of charge is not a source of ambiguities. It is well known that the net charges obtained by the Hückel method (when they are not zero) are overestimated with respect to SCF values [73]. In a large number of cases, the signs of the charges are the same; but for certain molecules, the signs of the net charges obtained by the two methods are in significant disagreement, as with azulene [74–65]. (see Tables 1 and 2).

Clearly, systematic reduction of the electronic charges obtained by the Hückel method would not suffice to give exact values.

### H. Examples

In Tables 1 and 2 some numerical results have been collected. These examples have been chosen among the very large number of charge calculations available in the literature in order to show the great variety of the molecules treated and of the methods employed.

Table 1. π-Methods (electronic π-charges)

Hückel [30]:

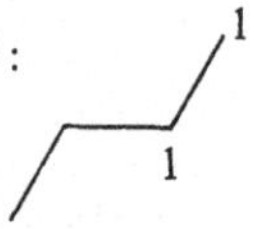


Butadiene

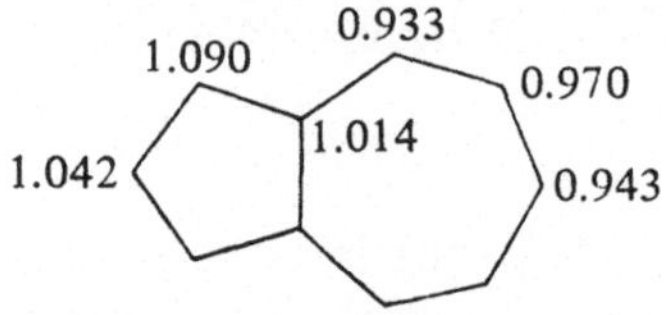


Azulene (Berthier Ref. [14])

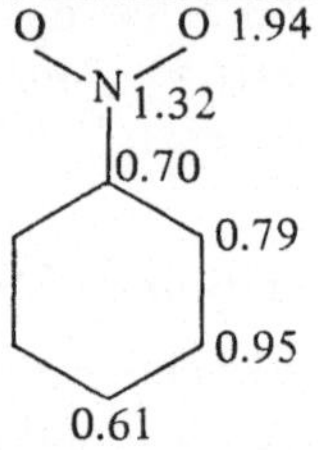


Nitrobenzene [75]

SCF [33]:

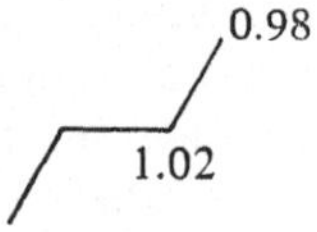


Butadiene [76]

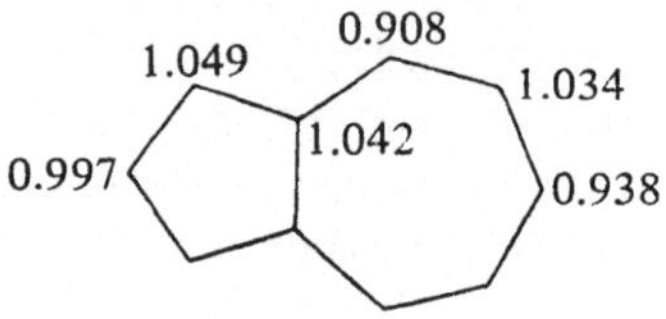


Azulene (Julg, Ref. [14])

Semiempirical methods:
-Pariser, Parr-type (semiempirical electronic integrals [77])

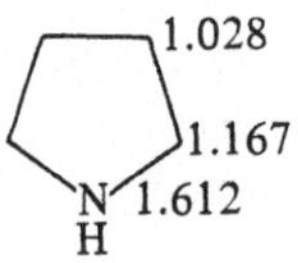


Pyrrole [78]

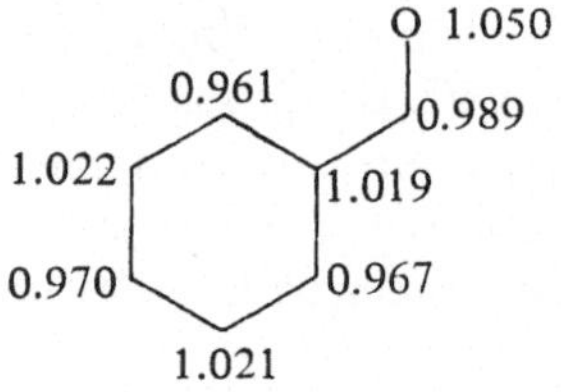


Benzaldehyde [79]

-LCAO-improved [56]:

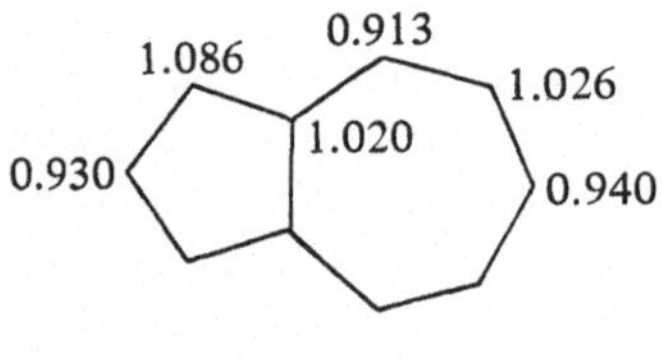


Azulene [80]

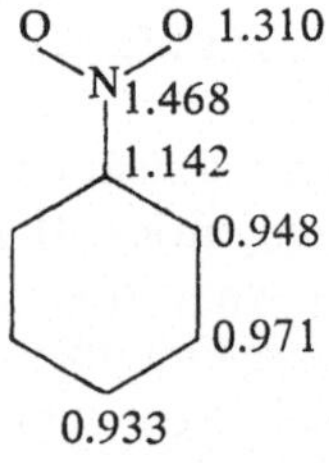


Nitrobenzene [81]

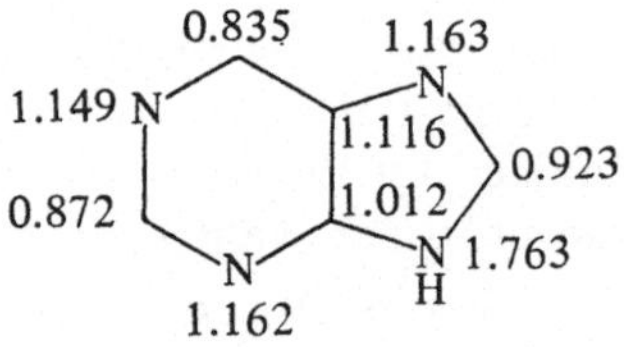


N(9)-purine [82]

-Variable electronegativity technique :

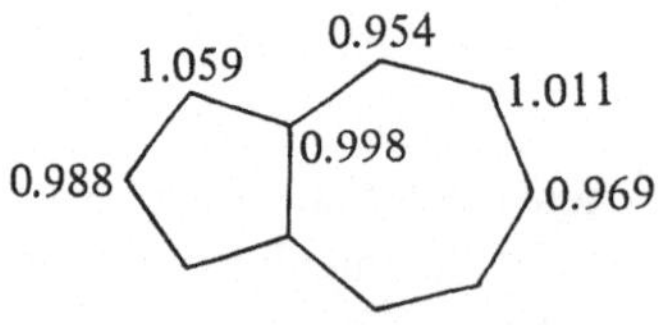


Azulene (Brown, Ref. [65])

For the same molecule there are sometimes very striking disagreements between different calculations: this is not necessarily the fault of the method, for it is quite possible that it may rather be due to the definition adopted for the charges. In fact, we have seen that the divergencies in the alkanes can be reduced by adopting a suitable definition. A general study of a sufficient number of examples is highly desirable in this context.

The examples have been classified by method: $\pi$ methods (Table 1), all-electron or valence-electron methods (Table 2): semiempirical and non-empirical

Table 2. All-electrons and valence-electrons methods

All-electrons method (SCF-ab-initio)

-0.2 0.4 -0.2
O–C–O

$CO_2$ [83]

$[Ni(CN)_4]^{2-}$ Net charges [84] { Ni : 0.46; C : -0.14; N : -0.47 }

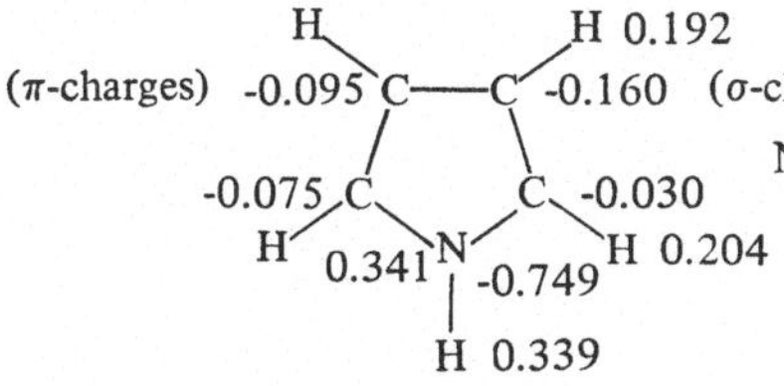


Net charges { N : -0.408; $C_1$, $C_4$ : -0.105; $C_2$, $C_3$ : -0.255 }

Pyrrole [85]

Semi-empirical methods :

-Extended Hückel [58]

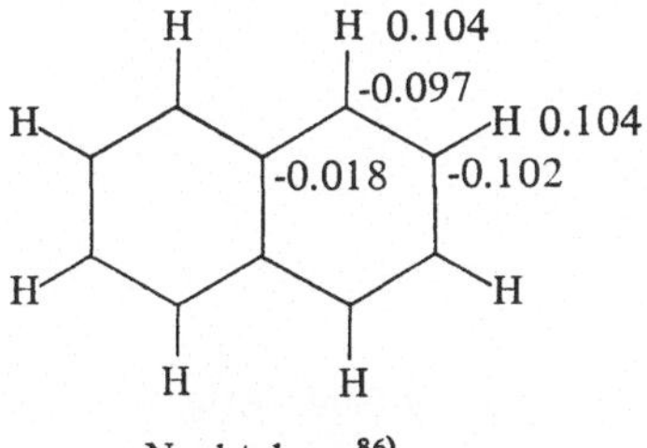


Naphtalene [86]
(net σ-charges)

$IF_7$ Net charges [87] { I : +2.930; F : -0.419 }

-CNDO (net charges)[88]

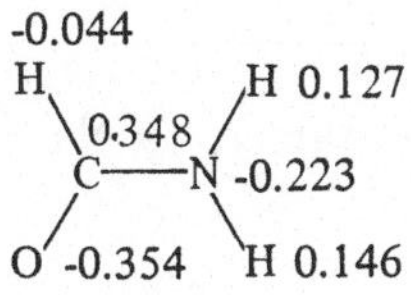


Formamide [89]

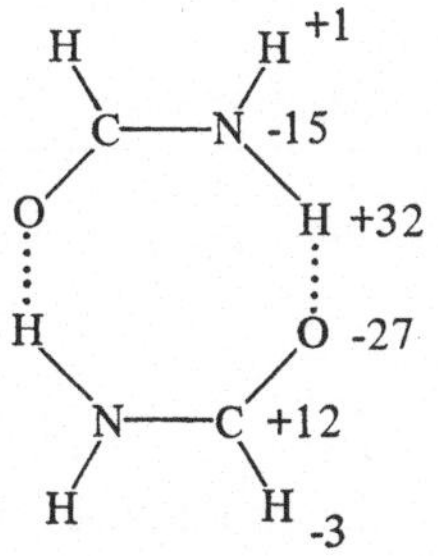


Dimere: variation/monomere
(in millielectron)

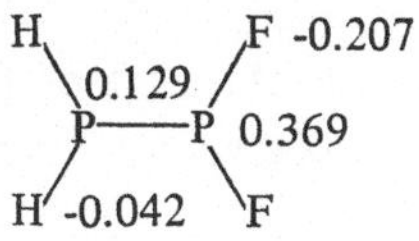


$H_2P–PF_2$ [90]

Table 2 (continued)

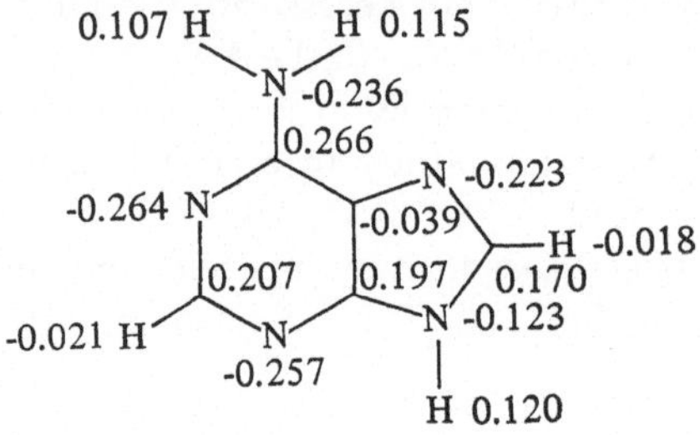


Adenine [91]

-Del Re's method[92] :

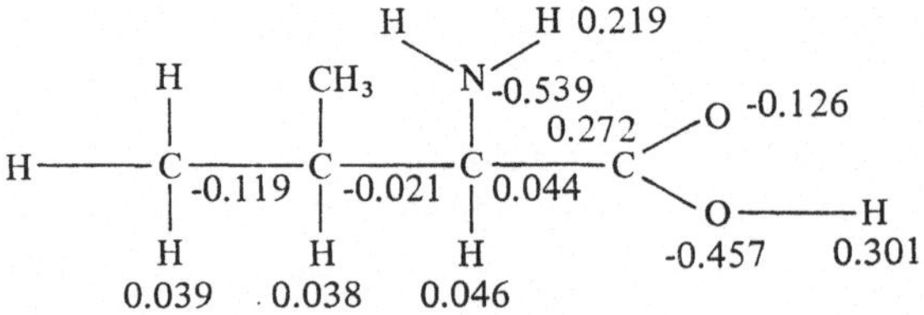


Valine [92]

(π-charges)

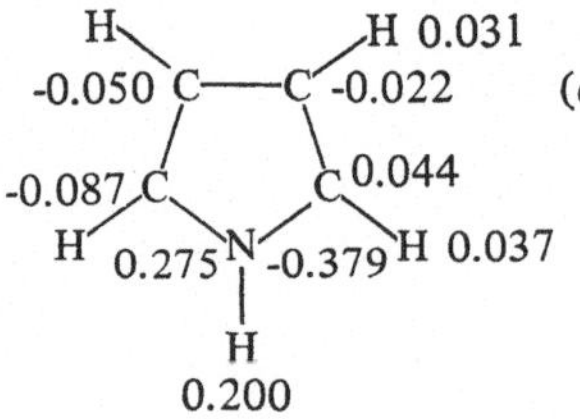


(σ-charges)

Net charges { N : -0.104; $C_1$, $C_4$ : -0.043; $C_2$, $C_3$ : -0.072 }

Pyrrole [93]

-PCILO [94] :

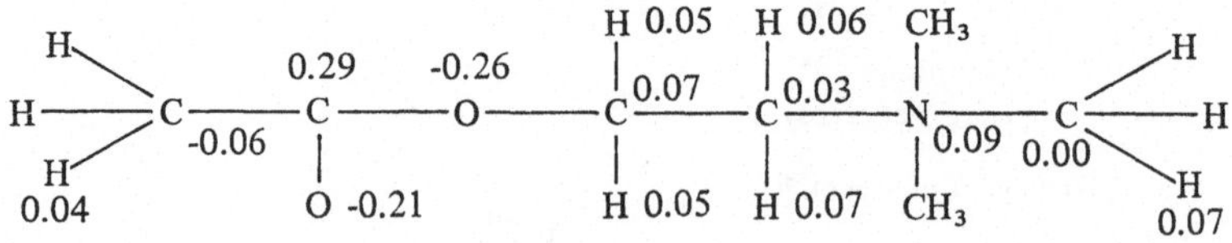


Acetylcholine [95]

methods. As in the rest of this article, only results obtained by the MO-LCAO method have been considered.

## I. Practical Conclusion

The above discussion shows that the problem of defining charges corresponding to the chemist's picture is far from solved in the general case. Apart from the problem of obtaining correct wave functions, only in the special case of $\pi$ orbitals can a valid definition be given, because the centers of gravity of the orbitals coincide with the nuclei, and no one nucleus lies in the region of maximum density of the others.

Should one then conclude that any attempt to evaluate charges from wave functions is hopeless in the general case? This is difficult to answer, but it seems certain that other directions should also be explored. At any rate, it appears that in spite of the difficulties encountered in attempts to define charges, the description obtained is more useful for the chemist than density maps directly drawn from wave functions: either maps of the densities as such or of the differential densities [97], or perspective views of three-dimensional maps [98].

## VII. Bond Moments

### A. Localization of Molecular Orbitals

At first sight, the description of a molecule as given by the method of molecular orbitals by linear combinations of atomic orbitals seems to contradict the classical idea of electron pairs localized between pairs of nuclei. In fact, owing to the way in which the function $\varphi_i$ is constructed, the electron density which corresponds to each pair, $2\varphi_i^2$, extends over the whole molecule. However, the various functions obtained by minimization of the energy have no absolute meaning. Without modification of the total wave function [Eq. (1)], one can indeed replace the set of the $\varphi$ functions corresponding to the doubly occupied levels by a set of functions $\varphi'$ which are orthonormal linear combinations of the $\varphi$ functions. This property results from the structure of the determinant used to describe the total wave function [99]. Taking advantage of the arbitrariness in the choice of this mathematical operation, the new functions $\varphi'$ are chosen so as to concentrate as far as possible the densities of the various pairs. Several mathematical criteria have been proposed, for example maximization of the sum of the repulsion energies of the two electrons (1 and 2) associated to the same spatial function [100].

$$\int\int \varphi'(1)^2 \frac{1}{r_{12}} \varphi'(2)^2 \, dv_1 \, dv_2$$

Others criteria have been proposed [101–102]. We shall not enter into the details of these methods, but give only the most important conclusions.

The first remark is that the inner shell electrons do not appear to differ from those of the corresponding free atoms. Bond formation only involves the outer shell electrons (valence electrons).

In saturated molecules, such as the alkanes, it is possible to localize to more than 95 % in regions between pairs of nuclei the densities which correspond to the different electron pairs. If the number of electron pairs of an atom is larger than the number of neighboring atoms, as in an amine, one of the pairs is localized in the vicinity of the corresponding nucleus but in a slightly excentric position.

In the case of unsaturated planar molecules, two types of localization are possible. The molecular wave functions are of two types: $\sigma$, symmetric with respect to the plane of the molecule, and $\pi$, antisymmetric with respect to that plane. The $\sigma$ and $\pi$ functions can be localized either separately or together. For instance, maximum localization for ethylene leads to the two schemes A and B of Fig. 5:

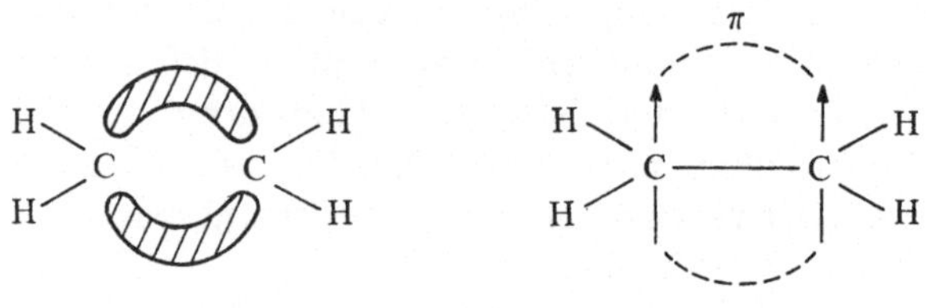


Fig. 5. The two types of localization of ethylene

The first procedure (A) gives two equivalent bent bonds between the carbon nuclei. The second procedure (B) gives one $\sigma$ and one $\pi$ bond.

The same situation takes place with linear molecules such as $N_2$ and $C_2H_2$. There are either three equivalent bent bonds or one $\sigma$ and two $\pi$ bonds [102].

The two schemes are obviously equivalent, but the second one is currently preferred, because it preserves the $\sigma$—$\pi$ separation which is important for planar conjugated molecules such as aromatic hydrocarbons. In the latter molecules there are a $\sigma$ frame formed by localized bonds and a delocalized $\pi$ system which must be considered in its entirety.

If the localization defect is neglected, the various molecular orbitals $\varphi'$ may be assumed to be constructed with pairs of linear combinations $t$ of orbitals centered on the nuclei between which the $\varphi'$ are localized: $\varphi'_{AB} \sim at_A + bt_B$. For a lone pair this will reduce to $\varphi' \sim t$. The combinations in question thus play the role of new basis orbitals, and are the so-called *hybrid* orbitals. They correspond to densities $t^2$ excentric with respect to the nuclei. To the extent to which the maximum of the density $\varphi'^2$ is situated on the internuclear axis, if orthogonality between hybrids belonging to the same nucleus is imposed, and if a minimal atomic orbital basis has been chosen, the mathematical structure of the hybrid orbitals is determined by the molecular structure [40–103]. Experimentally, the geometries of corresponding groups in related molecules are almost the same. This suggests that the hybrids are practically the same, and — as long as the effect of the environment on the given bond is roughly the same — so are the corresponding localized molecular functions. This remark explains the quasi-invariant character of the properties of the localized bonds of saturated molecules.

It is thus clear that orbital localization permits justification of the classical concepts of localized binding and lone electron pairs, as opposed to delocalized $\pi$ systems, as well as the notion of directed valency, a foundation of stereochemistry. Nevertheless, it must be noted that even if localization were perfect, the notion of bond electron pairs would be conventional [104]. In fact, if the descriptions in terms of the initial and the localized orbitals are fully equivalent in the sense that the total wave function is the same, quantities associated with a single pair of electrons are not invariant under the transformation. It has no absolute physical meaning. The same will hold for quantities associated with individual bonds.

## B. Bond Moments

The bond moment idea falls into the latter category of quantities. It was a very early idea of the chemist that each bond should be assigned a characteristic dipole moment, such that the total molecular electric dipole moment would be obtained by simple vector addition [105–19]. This idea is justified, as has been seen, in so far as perfect localization of molecular functions can be obtained. Now we have seen that the latter cannot be realized. Therefore, the partitioning of the total moment into partial bond moments will be merely an approximation. However, since the lack of localization is not particularly important, the error involved is not larger than the error tolerated on experimental dipole moments, say .1 to .2 Debyes. Therefore, in the following we shall neglect the lack of localization.

The dipole moment of a system can be defined if the system is electrically neutral, as has been seen in Section II-B. In order to define bond moments it will be necessary to partition the charges of the nuclei or of the cores in order to associate to them the electrons of individual bonds. In the case of a lone pair, carrying a $-2$ charge, it will be necessary to isolate a $+2$ charge on the corresponding nucleus.

Take for instance the methane molecule $CH_4$. The carbon nucleus, having a charge $+6$, is surrounded by its ls shell completely filled. It forms a core of charge $+4$ which, according to Gauss's theorem can be considered as a point charge. Let us divide this charge into four $+1$ charges.

By associating to each of these charges a proton of $+1$ charges with the pair of electrons having a $-2$ charge, localized between the carbon core and the proton under study, we have defined an electrically neutral system for which a dipole moment can be defined. We shall call the latter "moment of the CH bond" (Fig. 6).

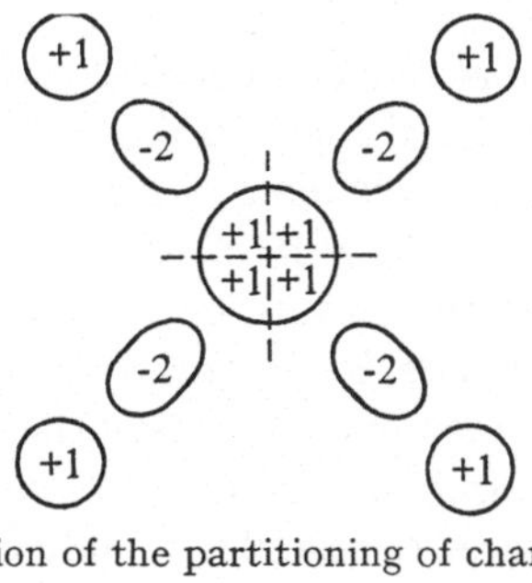


Fig. 6. Schematic representation of the partitioning of charges in the methane molecule

If a molecule has lone pairs, a similar procedure will apply. For instance, in the case of the ammonia molecule, the $+5$ charge of the nitrogen core will be partitioned into three $+1$ charges, each of which will be associated with a pair of electrons and a proton, and one $+2$ charge, to be associated with the lone pair (Fig. 7).

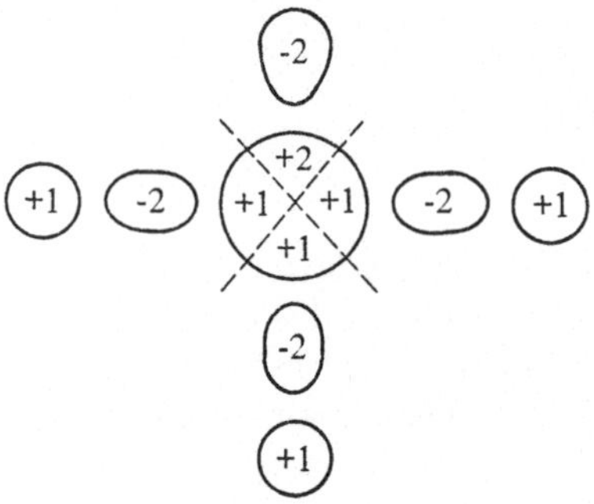


Fig. 7. Schematic representation of the partitioning of charges in the ammonia molecule

The moment of the NH bond and the moment of the lone pair are thus defined. The same holds for a molecule with a double bond, as $H_2CO$ (Fig. 8)

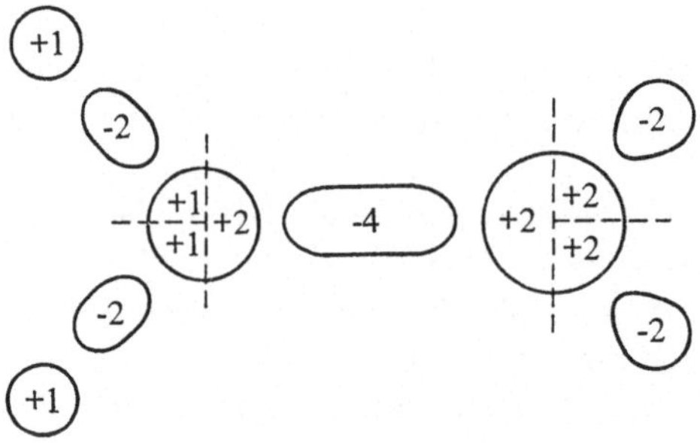


Fig. 8. Schematic representation of the partitioning of charges in the $H_2CO$ molecule. The two electron pairs of the double bond are taken together

## C. Case of Conjugated Planar Molecules

We have seen that in molecules one can consider two families of wave functions: the localizable $\sigma$ family and the non-localizable $\pi$ family. The dipole moment of the molecule is the sum of the dipole moments of the various $\sigma$ bonds ($\sigma$ moment) and of the $\pi$ moment. Of course, the $\pi$ moment could also be partitioned into bond contributions. But the values that would be obtained would not have the quasi-invariant character of the $\sigma$ bonds. For instance, in azulene the various CC bonds would have highly different moments. On the other hand, the partitioning into bond moments is of interest only when the moments obtained have a sufficient degree of *transferability*. We are interested here only in the $\sigma$ moment.

## D. Principle of Construction of a Table of Dipole Moments. The Problem of Lone Pairs

Knowing the dipole moments and the geometries of a sufficient number of molecules, and assuming that the bond moments are invariant, a trial and error procedure will permit construction of a table of standard bond moments. We shall not discuss here the details of this computation. We only point out that a difficulty of principle appears concerning lone pairs. They must be assigned dipole moments. This moment is easy to calculate with hybrid orbitals directed to the neighboring nuclei [103]. For instance, in the ammonia molecule, the structure of the hybrids can be determined from the $\widehat{HNH}$ angle. According to Ref. [106], the dipole moment corresponding to the lone pair is around 3.6. D. As the total moment of $HN_3$ is 1.5 D, one concludes that the moment of the NH bonds is 1.9 D, in the sense $N^+ - H^-$. Another way of looking at the problem consists in decomposing the total moment only along the three NH bonds. In this case, the bond moment obtained is 1.3 D, in the sense $N^- - H^+$. Of course, both points of view are valid for both reproduce the total moment, which is the only one having a meaning. However, the partitioning which takes the lone pair into account is more satisfactory from the theoretical point of view, for it follows more closely the path of electron-pair localization, even if the $N^+H^-$ polarity to which it leads may surprise the chemist. The problem of the halogens, which carry three lone pairs, is analogous. However, in practically all the halogen-containing molecules the halogen atom is linked to only one atom, and the presence of the lone pairs is permanent. Thus the moment corresponding to them is independent of the molecules but depends only on the neighboring atom (which is not the case for the lone pair carried by nitrogen, for its characteristics depend on the angles formed by the neighboring atoms with the nitrogen atom).

There is no major objection to including the lone pair moment in the carbon-halogen bond, which will as a result lie along the nuclear axis, with the halogen as the negative end.

## E. Moment of the CH Bonds

The moment of the CH bond is a special case deserving a separate discussion. The values to be assigned to the moments of all the other bonds in organic molecules depend on the choice made for it. For instance, the moment of the C—Cl bond will be derived from the moment of $CH_3Cl$ and of $CH_3CH_2Cl$; depending on the value chosen for the moment of the CH bond 1.56 [107] is obtained for CCl if $\mu(C^- - H^+) = 0.3$ D. The experimenters all agree on the fact that the CH moment is small, indeed not larger than a few tenths of a Debye; but they do not all agree on the polarity on the bond. The study of the chemical properties of hydrogen carbides suggests that hydrogen is the positive end in acetylene derivatives (hydrogen is substituted by metals, hydrogen bonds are formed with acetone [108–109], but forms the negative end in saturated hydrocarbons (carbocations are formed, hydrogen bonds are not). In ethylene derivatives, where hydrogen bonds are certainly very weak (ethylene is much less soluble in water and acetone than acetylene), and carbocations are not easily formed, the CH bond moment may be assumed to vanish. This trend is consistent with the electronegativity changes in the hybrid orbitals of carbon with which the localized bonds are formed [110]. With respect to the CH bond in ethylene, the CH bonds of acethylene and of ethane must have opposite polarities.

It seems reasonable to assign to the various CH bonds different polarities according to the types of compounds [111]:

$C^{+}_{sp^3}$ – $H^{-}$ : 0.3 D in saturated compounds;

$C_{sp^2}$ – H : 0 D in ethylene and benzene derivatives;

$C^{-}_{sp}$ – $H^{+}$ : 0.2 D in acetylene derivatives.

It would be interesting and even essential to define the dipole moments of these different types of bonds in a purely theoretical way. Unfortunately, as has been seen above, several reasons make this task impossible.

The different methods of calculation give, form Eq. (18), widely different but generally positive values for the net charges of the hydrogen atoms in the alkanes (see a more complete bibliography in Ref. [68]). Nevertheless, it seems that an Equation of the type Eq. (20) gives smaller differences and a weakly negative net charge [68].

### F. Induced Moments

Several arguments of a chemical [112] or of a theoretical nature [113] suggest that the presence of a strongly electronegative atom such as fluorine at one end of the chain of a linear hydrocarbon induces a general displacement of charges toward that atom. This effect is called *inductive effect*. Non-vanishing net charges appear on the chain atoms. Their signs are the same and their absolute values decrease with distance from the inducing atom. The effect is monotonic and damped.

This phenomenon is general. It appears in every bond between atoms substituted by atoms having different electronegativities. A modification of the bond moment results, and the new bond moment can be interpreted as the sum of the intrinsic bond moment and a moment induced by the neighboring atoms.

The importance of the induced moment depends on two factors: first, the electronegativity difference between the perturbing atom and the atom which carries it, second the bond polarizability, *i.e.* the ease with which the bond can be modified under the influence of an electric field. Although not negligible, induced moments are weak, so that in most cases they do not bring an important contribution to the total moment. However, for bonds whose intrinsic moment is small, their importance becomes considerable, the very sense of the moment being subject to inversion. For instance, in $H_3CCl$, even if the carbon atom is saturated ($sp^3$ hybridization), the attraction of chlorine on the electrons of the CH bonds may very well result in a $C^{-}$— $H^{+}$ polarity. Here again, the situation is not clear because of the inability of theory to give a value for the net charge of hydrogen.

### G. Note on Current Terminology

Consider a diatomic molecule or more generally a localized bond AB (for the general case, cf. Ref. [114]). Let $\varphi = c_A\chi_A + c_B\chi_B$ be the wave function describing the electron pair forming the bond.

Denote by $R_A$ and $R_B$ the coordinates of the nuclei, and by $R'_A$ and $R'_B$ those of the centers of gravity of the densities $\chi_A^2$ and $\chi_B^2$.

The dipole moment of the bond, as has been seen, may be written in the form:

$$\mu = - R_A - R_B + 2(c_A^2 R'_A + c_B^2 R'_B) + 4c_A c_B \int \chi_A z\, \chi_B\, dv$$

Let us write:

$$\int \chi_A z \chi_B dv = \frac{1}{2}\left(R'_A + R'_B + r\right) S_{AB}\,,\; R'_A = R_A + \Delta R_A \text{ and } R'_B = R_B + \Delta R_B$$

Then:

$$\mu = - R_A - R_B + 2(c_A^2 + c_A c_B\, S_{AB}) R_A + 2(c_B^2 + c_A c_B\, S_{AB}) R_B$$
$$+ 2(c_A^2 + c_A c_B\, S_{AB})\, \Delta R_A + 2(c_B^2 + c_A c_B\, S_{AB})\, \Delta R_B + 2 c_A c_B r\, S_{AB}\,.$$

If Eq. (17) is used for defining the electric charge, it follows:

$$\mu = (Q_A R_A + Q_B R_B) + (q_A \Delta R_A + q_B \Delta R_B) + 2 c_A c_B r\, S_{AB}$$

The total moment is thus the sum of three contributions [115]:

– the first term $(Q_A R_A + Q_B R_B)$ is called the *primary moment*. It corresponds to opposite net charges $Q_A$ and $Q_B$ located on the nuclei A and B.

– the second term $(q_A \Delta R_A + q_B \Delta R_B)$ originates from the fact that the centers of gravity of the distributions $\chi_A^2$ and $\chi_B^2$ do not coincide with the nuclei A and B. This is the case when the orbitals $\chi_A$ and $\chi_B$ are hybrids. This explains the name *hybridization moment* given to the terms $q_A \Delta R_A$ and $q_B \Delta R_B$. This name is not entirely correct if the term "moment" is limited to charge systems whose total charge is zero.

– the third term $2 c_A c_B r S_{AB}$, is called *overlap moment* because it contains the overlap integral. It is not a true moment in the sense specified above.

Two remarks are important in connection with the partitioning of the total moment. First, hybridization is a unitary transformation which does not change the total electron density of the orbitals if the latter are all equally occupied [103]. Therefore, the center of gravity of the whole set of hybrids is the nucleus: $\sum \Delta \boldsymbol{R}_i = 0$ It results that in a molecule, the sum of the hybridization moments associated with a given nucleus is practically zero if the latter does not carry any lone pair and if the $q$ charges are not very different [116] (*e.g.* in the carbon atoms). Second, the overlap moment depends on the definition chosen for the charges. If $\int \chi_A z \chi_B dv$ is partitioned according to Eq. (19) using Eq. (20) for the charge, that term disappears.

The above considerations show that not only is it not necessary, in general, to apply a partitioning of the bond moments, but that the danger exists of such a partitioning being wrongly interpreted so that more confusion results. In our opinion it is better to avoid it, use of total bond or lone pair moments as defined above normally being sufficient.

## VIII. Conclusion

From the discussion of the problems posed by the representation of molecules in terms of point charges or electric moments, it appears that, as in many other cases, rigor and simplicity are not easily reconciled. The chemical reactivity phenomena are extremely complicated and one cannot but admire the ingenuity of the past generations of chemists, who succeeded in building a consistent as well as productive structure by means of very simple concepts, such as bond pairs, electric charges, electron transfers, etc. The theories derived from quantum mechanics, on the other hand, give rise to complicated models which cannot be easily transcribed in familiar language without serious losses in rigor and are thus not easily used by the non-specialist. However, if some loss in rigor is accepted, the models correspond to a surprising extent with the models obtained by the chemists employing entirely different lines of thought. The loss in rigor involved is not as serious as may be thought, for the chemist is quite content with an approximate model. To him, what is important is to know that his model has a theoretical counterpart, even though the exact structure of the latter may be very complicated.

*Added in Proofs.* Combined X-ray and electron diffraction data can be used to analyze the electron density in a molecule. Identification of the square of the wave function with the electron density then gives the coefficients of the basis atomic orbitals in the various molecular orbitals, and therefore permits the evaluation of the electron populations. This method was used for instance with 1,3,5-trimethylbenzene, and good agreement with theoretical data was obtained [B. H. O'Connor, E. N. Maslen, Acta Cryst. *B 30*, 383 (1974)]. Of course, the results depend on the basis atomic orbitals and on the definition of the population.

*Acknowledgements.* The author is particularly grateful for the profitable discussions held with Prof. G. Del Re (Università di Napoli).

## IX. References

1) Davy, H.: Phil. Trans. 1807, 1.
2) Berzelius, J. J.: Essai sur la théorie des proportions chimiques et sur l'influence chimique de l'électricité. Paris 1819.
3) Dumas, J. B.: Compt. Rend. *6*, 633 (1838); Ann. Chim. Phys. *67*, 309 (1838).
4) Thomson, J. J.: Phil. Mag. *44*, 293 (1897). — Kaufmann, A.: Ann. Physik. *61*, 544 (1897).
5) Abegg, R.: Z. Anorg. Chem. *39*, 330 (1904).
6) Stieglitz, A.: Proc. Natl. Acad. Sci. U.S. *1*, 196 (1915); J. Am. Chem. Soc. *44*, 1293 (1922).
7) Noyes, W., Lyon, J.: J. Am. Chem. Soc. *23*, 463 (1901).
8) Lewis, G. N.: J. Am. Chem. Soc. *38*, 762 (1916).

9) Ingold, C. K.: Structure and mechanism in organic chemistry. Ithaca: Cornell Univ. Press 1953. — Remick, A. E.: Electronic interpretations of organic chemistry. New-York: John Wiley 1949. — Alexander, E. R.: Principles of organic reactions. New-York: John Wiley 1950.
10) Markovnikow, W.: Ann. *153*, 256 (1870).
11) Flürscheim, B.: J. Prakt. Chem. *66*, 321 (1902); *71*, 497 (1905); Chem. Ber. *39*, 3015 (1906). — Holleman, A. F.: Die direkte Einführung von Substituenten in den Benzolkern. Leipzig: Veit 1910.
12) Fry, H. S.: The electronic conception of valence and the constitution of Benzene. London—New-York: Longmans Green and Co. 1921; J. Am. Chem. Soc. *34*, 664 (1912); *36*, 248, 262, 1038 (1914); *37*, 855 (1915).
13) Anderson, A. G., Nelson, J. A.: J. Am. Chem. Soc. *72*, 3824 (1950).
14) Pullman, A., Berthier, G.: Compt. Rend. *227*, 677 (1948). — Berthier, G., Pullman, A.: Compt. Rend. *229*, 761 (1949). — Julg, A.: Compt. Rend. *239*, 1498 (1954).
15) Pople, J. A.: Proc. Phys. Soc. *A* LXVIII, 81 (1955). — Pariser, R.: J. Chem. Phys. *24*, 250 (1956). — Kolboe, S., Pullman, A.: Coll. international sur le calcul des fonctions d'onde moleculaires. Paris: C.N.R.S. 1958.
16) Wheland, G. W.: J. Am. Chem. Soc. *64*, 900 (1942). — Pullman, B., Pullman, A.: Les théories électroniques de la chimie organique. Paris: Masson 1952.
17) Julg, A., Carles, P.: J. Chim. Phys. *59*, 852 (1962). — Scrocco, E., Tomasi, J.: In: Topics in current chemistry *42*, p. 130. Berlin—Heidelberg—New York: Springer 1973.
18) Kharasch, M. S., Mayo, F. R.: J. Am. Chem. Soc. *55*, 2468 (1933).
19) see: Le Fèvre, J. W.: Dipole moments. London: Methuen 1953. — Barriol, J.: Les moments dipolaires. Paris: Gauthier-Villars 1957.
20) Onsager, L.: J. Am. Chem. Soc. *58*, 1486 (1936).
21) Debye, P.: Physik. Z. *13*, 97 (1912).
22) Eisenschitz, R., London, F.: Z. Physik *60*, 491 (1930). — London, F.: Z. Physik *63*, 245 (1930).
23) see: Messiah, A.: Quantum mechanics. Amsterdam: North-Holland 1964. — Landau, L., Lifschitz, E.: Quantum mechanics, Moscow: Mir. 1966.
24) Schrödinger, E.: Ann. Physik. *79*, 361, 478; *80*, 437; *81*, 109 (1926).
25) Born, M., Oppenheimer, J. R.: Ann. Physik *84*, 457 (1927).
26) Heitler, W., London, F.: Z. Physik *44*, 455 (1927).
27) see Richards, W. G., Walker, T. E. H., Hinkley, R. K.: A bibliography of *ab initio* molecular wave functions. Oxford: Clarendon Press 1971.
28) Slater, J. C.: Phys. Rev. *34*, 1293 (1929).
29) Roothaan, C. C. J.: Rev. Mod. Phys. *32*, 179 (1960).
30) Hückel, E.: Z. Physik *70*, 204 (1931); *72*, 310 (1931); *76*, 628 (1932).
31) Hall, G. G.: Repts. Progr. Phys. *22*, 1 (1959). — Kutzelnigg, W., Del Re, G., Berthier, G.: Fortschr. Chem. Forsch. *22*, 86 (1971). — Bonyard, K. E., Sutton, A.: J. Phys. *B*, *5*, 773 (1972).
32) see: Scrocco, E., Tomasi, J.: Ref. [17], p. 136.
33) Roothaan, C. C. J.: Rev. Mod. Phys. *23*, 69 (1951).
34) Scrocco, E., Tomasi, J.: Ref. [17].
35) Pack, G. R., Wang, H., Rein, R.: Chem. Phys. Letters *17*, 381 (1972). — Riera, A., Meath, W. J.: Mol. Phys. *24*, 1407 (1972). — Bonaccorsi, R., Cimiraglia, R., Scrocco, E., Tomasi, J.: Theoret. Chim. Acta *33*, 97 (1974).
36) Bonaccorsi, R., Petrongolo, C., Scrocco, E., Tomasi, J.: Theoret. Chim. Acta *20*, 331 (1971).
37) Born, M., Mayer, J. E.: Z. Physik *75*, 1 (1932).
38) for a simple method of calculation, see: Evjen, H. M.: Phys. Rev. *39*, 675 (1932). — Höjendahl, K.: Kgl. Danske Vid. Selskab Math. Medd. *16*, 133 (1938).
39) Jenkins, H. D. B., Waddington, T. C.: J. Inorg. Mol. Chem. *34*, 2465 (1972).
40) Van Vleck, J. H., Sherman, A.: Rev. Mod. Phys. *7*, 174 (1935).
41) Julg, A., Julg, O.: Exercices de chimie quantique, p. 37. Paris: Dunod 1967.
42) Frost, A. A., Prentice, B. H., Rouse, R. A.: J. Am. Chem. Soc. *89*, 3064 (1967).

43) Platt, J. R.: Handbuch der Physik *37 bis*, 173 (1961). — Berthier, G.: In: Aspects de la chimie quantique contemporaine, p. 61. Paris: C.N.R.S. 1971. — André, J. M., André, M. C., Delhalle, J., Leroy, G.: Ann. Soc. Sci. Bruxelles *86*, 130 (1972).
44) Chirgwin, B. H., Coulson, C. A.: Proc. Roy. Soc. (London) *A 201*, 197 (1950).
45) Löwdin, P. O.: J. Chem. Phys. *18*, 365 (1950). — Daudel, R., Laforgue, A.: Compt. Rend. *233*, 623 (1951).
46) Béry, J. C.: Theoret. Chim. Acta *7*, 249 (1967).
47) Cizek, J.: Mol. Phys. *6*, 19 (1963).
48) Jug, K.: Theoret. Chim. Acta *31*, 63 (1973).
49) Fenske, R. F., Caulton, K. G., Radtke, D. D., Sweeney, C. C.: Inorg. Chem. *5*, 951 (1966); *5*, 960 (1966). — Barnett, G. P., Pires Costa, M. C., Ferreira, R.: Chem. Phys. Letters *25*, 351 (1974).
50) Mulliken, R. S.: J. Chem. Phys. *23*, 1833, 1841, 2338, 2343 (1955).
51) Berthier, G., Messer, A. Y., Praud, L.: In: The Jerusalem symposia on quantum chemistry and biochemistry, vol. VIII, p. 174. New-York: Academic Press 1971.
52) Ros, P., Schmit, G. C. A.: Theor. Chim. Acta *4*, 1 (1966).
53) Bourdeaux, E. A.: *in* Stout, E. W., Politzer, P.: Theoret. Chim. Acta *12*, 379 (1968). — Cusachs, L. C.: *ibidem.*
54) Landshoff, R.: Z. Physik *102*, 201 (1936).
55) Löwdin, P. O.: J. Chem. Phys. *18*, 365 (1950).
56) Julg, A.: J. Chim. Phys. *57*, 19 (1960).
57) Del Re, G.: Intern. J. Quant. Chem. *1*, 293 (1967).
58) Hoffmann, R.: J. Chem. Phys. *39*, 1397 (1963).
59) Coulson, C. A., Longuet-Higgins, H. C.: Proc. Roy. Soc. (London) *A 191*, 39 (1947).
60) Julg, A.: Tetrahedron *19*, suppl. *2*, 25 (1963).
61) Politzer, P., Harris, R. H.: J. Am. Chem. Soc. *92*, 6451 (1970).
62) Hartman, H., Jug, K.: Theoret. Chim. Acta *3*, 439 (1965).
63) Hinze, J., Whitehead, M. A., Jaffé, H. H.: J. Am. Chem. Soc. *85*, 184 (1963).
64) Schaefer, T., Schneider, W. G.: Can. J. Chem. *41*, 966 (1963).
65) Julg, A.: J. Chem. Phys. *52*, 377 (1955). — Pariser, R.: J. Chem. Phys. *25*, 1112 (1956). — Brown, R. D., Heffernan, M. L.: Australian J. Chem. *13*, 38 (1960).
66) Del Re, G., Pullman, B., Yonegawa, T.: Biochim. Biophys. Acta *75*, 153 (1963).
67) see the bibliography in Ref. [68].
68) Fliszàr, S., Kean, G., Macaulay, R.: J. Am. Chem. Soc. *96*, 4353 (1974). — Fliszàr, S., Goursot, A., Dugas, H.: J. Am. Chem. Soc. *96*, 4358 (1974).
69) Thomas, T. D.: J. Chem. Phys. *52*, 1373 (1970).
70) Henry, H., Fliszàr, S.: Can. J. Chem. *52*, 3799 (1974).
71) Jørgensen, Ch. K., Berthon, H., Balsenc, L.: J. Fluorine Chem. *1*, 327 (1971).
72) see a good bibliography in Ref. [68].
73) Berthier, G.: J. Chim. Phys. *50*, 344 (1953).
74) Coulson, C. A., Longuet-Higgins, H. C.: Rev. Sci. *85*, 929 (1947). — Berthier, G., Pullman, A.: Ref. [14].
75) Higasi, K., Baba, H., Rembaum, A.: Quantum organic chemistry, p. 7. New York: Interscience Publishers 1965.
76) Parr, R. G., Mulliken, R. S.: J. Chem. Phys. *18*, 1338 (1950).
77) Pariser, R., Parr, R. G.: J. Chem. Phys. *21*, 466, 767 (1953). — Paoloni, L.: Nuovo cimento *4*, 410 (1956). — Mataga, N., Nishimoto, K.: Z. Physik. Chem. (Frankfurt) *13*, 140 (1957).
78) Dahl, J. P., Hansen, A. E.: Theoret. Chim. Acta *1*, 199 (1963).
79) McClelland, B. J.: Trans. Faraday Soc. *468*, 2073 (1961).
80) François, Ph., Julg, A.: Theoret. Chim. Acta *11*, 128 (1968).
81) Bonnet, M.: Theoret. Chim. Acta, *11*, 361 (1968).
82) Julg, A., Carles, P.: Theoret. Chim. Acta *17*, 30 (1970).
83) Mulligan, J. F.: J. Chem. Phys. *19*, 347 (1951).
84) Demuynck, J., Veillard, A.: Theoret. Chim. Acta, *28*, 241 (1973).
85) Clementi, E., Clementi, H., Davis, D. R.: J. Chem. Phys. *46*, 4725 (1967).
86) Fukui, G. K.: In: Modern quantum chemistry (Istanbul Lectures), ed. by O. Sinanoglu, Part I, p. 58. New York: Academic Press 1965.

87) Oakland, R. L., Duffey, G. H.: J. Chem. Phys. *46*, 19 (1967).
88) Pople, J. A., Segal, G. A.: J. Chem. Phys. *43*, 5136 (1965).
89) Pullman, A., Berthod, H.: Theoret. Chim. Acta *10*, 461 (1968).
90) Bach, M. C., Crasnier, F., Labarre, J. F., Leibovici, Cl.: J. Mol. Struct. *13*, 171 (1972).
91) Giessner-Prettre, C., Pullman, A.: Theoret. Chim. Acta *9*, 279 (1968).
92) Del Re, G.: J. Chem. Soc. *1958*, 4031. In: Electronic aspects of biochemistry, ed. by B. Pullman, p. 221. New York: Academic Press 1964.
93) Momicchioli, F., Del Re, G.: J. Chem. Soc. *B*, 674 (1969).
94) Diner, S., Malrien, J. P., Claverie, P.: Theoret. Chim. Acta *13*, 1 (1968).
95) Pullman, B., Courrière, Ph.: Theoret. Chim. Acta *31*, 1 (1973).
96) Wahl, A. C.: Science *151*, 961 (1966).
97) Roux, M.: J. Chim. Phys. *55*, 754 (1958); *57*, 53 (1960).
98) Beebe, N. H. F., Sabin, J. R.: Chem. Phys. Letters *24*, 389 (1974).
99) Fock, V.: Z. Physik *61*, 126 (1930).
100) Lennard-Jones, J. E., Pople, J. A.: Proc. Roy. Soc. (London) *A 202*, 166 (1950). — Edmiston, C., Ruedenberg, K.: Rev. Mod. Phys. *35*, 457 (1963).
101) Del Re, G., Esposito, U., Carpentieri, M.: Theoret. Chim. Acta *6*, 36 (1966).
102) Ruedenberg, K.: In: Modern quantum chemistry (Istanbul Lectures), ed. by O. Sinanoglu, Part I, p. 85. New York: Academic Press 1965.
103) Julg, A.: Chimie théorique, p. 29. Paris: Dunod 1964. — Julg, A., Julg, O.: Ref. [41], p. 17.
104) Löwdin, P. O.: Advan. Phys. (Phil. Mag. Suppl.) *5*, 12 (1956).
105) Thomson, J. J.: Phil. Mag. *46*, 407 (1923). — Williams, J. W.: Phys. Z. *30*, 391 (1929).
106) Burnelle, L., Coulson, C. A.: Trans. Faraday Soc. *53*, 403 (1957). — Julg, A., Bonnet, M.: J. Chim. Phys. *60*, 742 (1963).
107) Smyth, C. P.: J. Phys. Chem. *41*, 209 (1937); J. Am. Chem. Soc. *60*, 183 (1938).
108) Gent, W. L. G.: Quater, Rev. *2*, 383 (1948).
109) Julg, A.: Chimie théorique, *loc. cit.* Ref. [103], p. 216.
110) Julg, A., Julg, O.: Ref. [41], p. 36.
111) Coulson, C. A.: Trans. Faraday Soc. *38*, 433 (1942). — Walsh, A. D.: Faraday Soc. Disc. *2*, 18 (1941).
112) Lewis, G. N.: Valence and the structure of atoms and molecules, p. 139. New York: The Chemical Catalog. Co. 1923. — Lucas, H. J.: J. Am. Chem. Soc. *48*, 1827 (1926). — Ingold, C. K.: Chem. Rev. *15*, 225 (1934). — Groves, L. G., Sugden, T. M.: J. Chem. Soc. 1992 (1937).
113) Sandorfy, C.: Can. J. Chem. *33*, 1337 (1955). — Julg, A.: J. Chim. Phys. *53*, 548 (1956).
114) Rinaldi, D., Rivail, J. L., Barriol, J.: Theoret. Chim. Acta *22*, 298 (1971).
115) Mulliken, R. S.: J. Chim. Phys. *46*, 497, 675 (1949).
116) Del Re, G.: Intern. J. Quant. Chem. *1*, 293 (1967).

Received January 9, 1975

# CRAMS
# An Automatic Chemical Reaction Analysis and Modeling System[1,2]

**Robert S. Butler[3] and Paul A. D. deMaine[4]**

Computer Science Department, the Pennsylvania State University, University Park, Pennsylvania 16802 U. S. A,

## Contents



---

[1] Taken from the Ph. D. Thesis by R. S. Butler, The Pennsylvania State University, 1974.

[2] Supported in part by funds from the National Science Foundation (Grant No. GJ-42336X).

[3] Post-Doctoral Fellow at The Institute for Computing Sciences and Technology, National Bureau of Standards, Gaithersburg, Maryland.

[4] Senior U. S. Scientist Award of the Alexander von Humboldt Foundation at Technische Universität, Munich, Germany.

# 1. Introduction

## 1.1. Statement of the Problem

A reaction system is a physical phenomenon whereby one or more reactants interact in any of several different modes over a period of time. The rate at which a reaction proceeds is a function of the concentrations of the reactants. Each reactant in the system is represented by a differential equation. If equilibrium or steady state assumptions are assumed, algebraic equations may be introduced, and the original equations may be altered to reflect these assumptions.

Thus, the reaction model is represented by a system of differential and algebraic equations. Although they may be linear, in the general case they are nonlinear. The parameters of the model are of two different types: constant and variable. A constant parameter may be either known or unknown, and may appear as a rate or equilibrium constant or within a rate expression. A variable parameter is the concentration of a reactant. A variable parameter is known if sufficient data are given to describe its behavior over the period of time of the reaction.

Typical questions concerning the parameters of the model are:

1. Given values for some of the parameters, which other parameters can be computed?

2. How can these other parameters be computed?

3. For what parameters is information needed to compute values for other desired parameters?

4. Given values for some of the parameters and their associated experimental errors, what can be said concerning the mathematical validity of the proposed reaction model?

This paper describes an operational system called CRAMS that automatically answers such questions.

The specific types of reactions discussed are chemical reactions in which the reactants may be compounds, fragments of compounds, elements, or ions; the stoichiometric units are moles; and the concentrations of the reactants are given in terms of moles/liter. However, this does not exclude the use of the system by other scientific disciplines. For example, Garfinkel and Sack [1] describe an application of reaction system models to ecological systems. The medium in which the reactions take place may be a pond or a forest and the reactants may be particular plant or animal species. In this case the reactions describe the growth and decay of species in the pond. The concentration of a reactant is more conveniently thought of as the population of a particular plant or animal species. In comparison with chemical reactions, both rate and equilibrium reactions occur in each type of model, but the rate expressions are considerably different. As certain conventions apply to chemistry that may not apply to other areas, the model is usually discussed in terms of chemical reactions. For example, the rate of a chemical reaction is usually proportional to the product of concentrations of reactants on the left hand side of the reaction.

To illustrate the questions that might be asked by a chemist, consider the following simplified version of the mechanism of four-component condensation for Peptide Synthesis (proposed in Schemes XVI and XVIa by Ugi [39]). It should be noted that stereoisomeric intermediates have not been identified and that a detailed study of both schemes will be offered in a future paper.

$$A \underset{K_1}{\overset{H^+}{\rightleftharpoons}} X_1 \underset{K_2}{\overset{X_5}{\rightleftharpoons}} X_2 \underset{K_3}{\overset{H^+}{\rightleftharpoons}} X_3 \underset{K_4}{\overset{X_5}{\rightleftharpoons}} X_4$$

$$C \bigg| k_1 \qquad\qquad C \diagdown k_2 \;\; C \bigg| k_3 \;\; C \diagup k_4$$

$$N \xrightarrow[k_5]{X_5} Z \xrightarrow[k_6]{} Y$$

$$B \underset{K_0}{\rightleftharpoons} H^+ + X_5$$

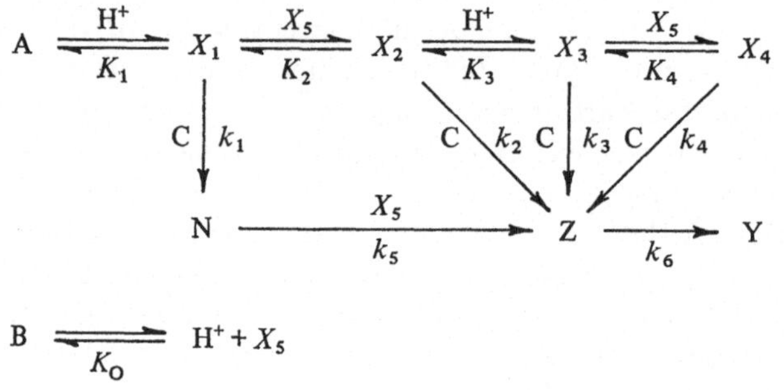

Here $K_i$ and $k_i$ are equilibrium and rate constants respectively; and $A \underset{K_1}{\overset{H+}{\rightleftharpoons}} X_1$ means $A + H^+ \underset{K_1}{\overset{}{\rightleftharpoons}} X_1$. A, B, and C are the starting materials. $X_1$, $X_2$, $X_3$, $X_4$, N and Z are complex stereoisomeric intermediates explicitly defined in Scheme XVI of [39]. Y is a mixture of the (S, S)- and (R, R)-valine derivatives. The starting materials are isobutyraldehyde-(S)-$\alpha$-phenylethylimine (A); benzoic acid (B), and t-butylisocyanide (C). $X_5$ is the benzoate ion, $C_6H_3COO^-$. It should be noted that none of the rate steps occur until t-butyl isocyanide (C) is added and that, at a constant temperature, special conditions are obtained by varying the initial concentration of A, B, C, $H^+$ or $X_5$. Parameters are equilibrium constants, rate constants, and concentration-time data for all the chemical compounds or intermediates that are involved.

Questions that might be asked are:

(i) For what combinations of parameters must values be given so that the CRAMS system can compute values for all unknown parameters?

(ii) What information is needed by CRAMS so that it can compute concentration-time values for Z and $k_1$?

(iii) If $k_5$ and $k_6$ are very large (*i.e.* the reactions are fast) and data for $H^+$, $X_5$ and C are given, what parameters can be computed?

## 1.2. Previous Related Studies

Most of the previous work on reaction systems has been directed specifically to biochemical systems. A comprehensive review of the subject is given by Garfinkel *et al.* [2]. Most of the methods discussed were either simulation systems or curve fitting systems.

Simulation may be defined as the generation of concentration data given initial concentrations and values for all of the constant parameters. Concentration data is generated by simultaneously solving the differential equations associated with each of the reactants. The disadvantage of this restricted approach is that all constant parameters must be given, even if some cannot be measured. Also, in mock simulation systems there is generally no method for directly using measured concentration data. Several major systems exist which translate a description of a reaction model into differential equations and solve the resulting equations. DeTar [3] describes a computer program which allows both rate and equilibrium equations to be specified. Garfinkel [4] has written a simulation program for large complex biochemical systems. The program of Chance and Shepard [5] allows representation of the system in terms of both differential rate equations and algebraic rate laws. Rate law methods allow the simplification of the equations which represent the reaction system through certain assumptions about the behavior of intermediate reactants. Other workers have also used rate law methods to increase the speed of simulation [6, 7]. Groner *et al.* [8] have written a program which translates the description of a reaction model into CSMP (Continuous System Modeling Program) statements, and CSMP subsequently carries out the simulation. Curtis and Chance [9] and Chandler *et al.* [10] have developed techniques whereby rate constants are automatically readjusted as repeated simulations are performed. In addition to the general disadvantages of this "simulation only" approach, none of these systems has any facility for automatically reducing the number of differential equations that must be solved to evaluate the reaction system.

The curve fitting approach is in contrast to the simulation approach. Curve fitting systems solve for the constant parameters if values for all of the variable parameters are given and the time scales for the different variable parameters is the same. The disadvantage of this restricted approach is that it is unlikely that all of the variable parameters can be measured experimentally. Moreover, in those few cases for which such data exist, the time scales are not exactly the same and certain simplifying assumptions have to be made. Several researchers have written general programs to calculate rate constants directly. deMaine [11] uses non-statistical curve fitting techniques to calculate the rate constants if concentration data for all reactants are given. The SAAM program by Berman *et al.* [12] uses statistical methods to choose from a variety of models that one which fits the experimental data best. Cleland [13, 14] has also used statistical methods in processing experimental data to elucidate several enzyme mechanisms. Pring [15–17] describes a system which uses nonlinear regression techiques to analyze reaction systems. Swann [18] surveys nonlinear optimization techniques used by biochemists. Kowalik and Morrison [19], Arihood and Trowbridge [20] and Atkins [21] have also written programs using nonlinear regression to analyze reaction models.

Several programs have been written specifically for a very restricted class of "equilibrium only" problems. The Pit Method of Sillen and Warnquist [22] has been widely used to solve for equilibrium constants in inorganic systems that have one or more simultaneous reversible reactions. DeLand [23] uses goal-seeking routines to facilitate the matching of data, but free energy data for all reactants is required. Bos and Meershoek [24] have written a PL/1 program which uses the Newton-Raphson iteration to compute equilibrium constants in complex systems.

In summary, none of these systems, or combinations of any of the systems that have been described in the literature, solve the general problems presented in Section 1.1. There appears to exist no general method which can determine which parameters are necessary to solve for the other unknown parameters in other than very elementary cases. In addition, none of the methods appears to consider the problem of mathematical model validation, *i. e.*, no analytical check is made to insure that the equations are valid algebraic descriptions of the data that is used to test the model.

The new system, CRAMS, that is described in the remainder of this paper automatically solves the general problems presented in Section 1.1. Details of its implementation [25a)] and operation [25b)] will be found elsewhere.

## 2. Overall Design of the CRAMS System

### 2.1. Overview

The **C**hemical **R**eaction **A**nalysis and **M**odeling **S**ystem, CRAMS, described in this paper, is a computer program that can automatically solve many of the problems presented in Section 1.1. The two major problems that are solved by CRAMS may be identified as the computing problem and the predictor problem.

The computing problem is concerned with calculating the maximum number of unknown parameters of a proposed reaction system from available experimental data. This data can be any combination of values for constant parameters (rate and equilibrium constants) and variable parameters (concentration versus time data). Moreover, data for different variable parameters need not have the same time scale. When the unknown parameters are calculated, it is important that the mathematical validity of the proposed model be determined in terms of the experimental accuracy of the data. Also, if it is impossible to solve for all unknown parameters, then the model must be automatically reduced to a form that contains only solvable parameters. Thus, the input to CRAMS consists of: 1) a description of a proposed reaction system model and, 2) experimental data for those parameters that were measured or previously determined. The output of CRAMS is: 1) information concerning the mathematical validity of the model and 2) values for the maximum number of computable unknown parameters and, if possible, the associated reliabilities. The system checks for model validity only in those reactions with unknown rate constants. Thus a simulation-only problem does not invoke any model validation procedures.

The predictor problem is concerned with determining which additional data are needed to compute any subset or all of the remaining unknown parameters. In a predictor type problem, the input to the CRAMS consists of: (1) a description of a proposed reaction system model, (2) a list of parameters that have been experimentally measured or previously determined, and (3) a list of parameters which have not been previously determined and which cannot be measured experimentally. The output of CRAMS consists of a list of various minimal combinations of additional parameters needed to calculate some or all of the remaining unknown parameters. It should be noted that the predictor problem is not discussed in the literature, presumably because of the general trend towards simulation only, or curve fitting only, systems. Except in the simulation only or curve fitting only cases, the only known existing system capable of solving the computing type problems associated with predictor output from CRAMS is CRAMS itself.

It must be emphasized that it is not possible for the system to choose from a selection of models. In both the computing and predictor type problems, CRAMS is capable of working only with one preselected reaction system model at a time. However, the user can test a number of different models in search of one which best describes his experimental data.

## 2.2. Input to the System

In this section the permitted types of reaction system models are precisely defined. The simple, free format input language that has been developed to describe the model and its associated experimental data to the computer program is defined elsewhere [25b)]. As this section is concerned with models that can be handled by CRAMS, some definitions differ slightly from those given in Section 1.1., where a more general model is discussed.

A reaction system consists of $NRCT$ reactants. Four vectors, all of length $NRCT$, are associated with these reactants. $X(i)$ is the label of the $i$th reactant. $IC(i)$ is the initial concentration of the $i$th reactant. The system assumes that the initial concentrations for all reactants are known, and unless specified by the user, the initial concentration for a reactant defaults to zero. $C(i)$ is the current value of the concentration of the $i$th reactant and $CP(i)$ is the current value of the derivative with respect to time of the concentration of the $i$th reactant.

The $NRCT$ reactants react in any of $NRXN$ reactions of two types: $NRK$ rate reactions and $NEK$ equilibrium reactions. For $1 \leq j \leq NRK$, the $j$th reaction (a rate reaction) is given by

$$M(1,j)*X(1) + \ldots + M(NRCT,j)*X(NRCT) \longrightarrow$$
$$N(1,j)*X(1) + \ldots + N(NRCT,j)*X(NRCT) \;.$$

This reaction means that $M(i,j)$ units of $X(i)$ for all $1 \leq i \leq NRCT$ react to form $N(k,j)$ units of $X(k)$ for all $1 \leq k \leq NRCT$. The $M(i,j)$'s and $N(i,j)$'s are stoichiometric coefficients and may be any real numbers. For $NRK < j \leq NRXN$ the $j$th reaction (an equilibrium reaction) is given by:

$$M(1,j)*X(1) + \ldots + M(NRCT,j)*X(NRCT) \leftrightharpoons$$
$$N(1,j)*X(1) + \ldots + N(NRCT,j)*X(NRCT) \;.$$

The meaning of this reaction is the same as that of a rate reaction, except that here the reverse reaction also occurs.

The $FLUX$ matrix is a two dimensional $NRCT$ by $NRXN$ matrix such that $FLUX\ (i,j) = N(i,j) - M(i,j)$ with $1 \leq i \leq NRCT$ and $1 \leq j \leq NRXN$. Notice that the $FLUX$ matrix is a convenient representation of the stoichiometric coefficients of the total reaction system. Each row of the matrix represents one of the $NRCT$ reactants and each column of the matrix represents one of the $NRXN$ reactions.

Two vectors, each of length *NRXN*, are associated with each reaction. $K(j)$ is the actual numerical value of the rate or equilibrium constant associated with the $j$th reaction. If $j \leq NRK$, then $K(j)$ is a rate constant. Otherwise, $K(j)$ is an equilibrium constant. $R(j)$ is the current value of the rate or equilibrium expression associated with the $j$th reaction. If $j \leq NRK$, then $R(j)$ is a rate expression. Otherwise, $R(j)$ is an equilibrium expression.

In the case where reaction $j$ is a rate reaction, $R(j)$ is restricted to being a continuous function of the current values of the concentrations of reactants. This restriction is made because it simplifies the formulation of a solvable set of equations. The default rate expression for the $j$th reaction is given by:

$$R(j) = \prod_{i=1}^{NRCT} C(i)^{**}M(i,j) \, .$$

However, other rate expressions may be specified to allow for such special cases as catalyzed or inhibited reactions. This also permits rate expressions derived by rate law methods to be specified explicitly, even though there is no facility in CRAMS for using rate laws automatically. The rate of the $j$th reaction is given by $K(j)^{*}R(j)$.

In the case where reaction $j$ is an equilibrium reaction,

$$K(j) = R(j) = \frac{\prod_{i=1}^{NRCT} C(i)^{**}N(i,j)}{\prod_{i=1}^{NRCT} C(i)^{**}M(i,j)} \, .$$

Thus, in the equilibrium case, the user may not specify the equilibrium expression.

The allowable types of unknown parameters are: (1) concentrations of reactants, (2) rate constants, and (3) equilibrium constants. A vector *KP* of length $NPAR = NRCT + NRXN$ contains the current status of each of these parameters. The vector *KP* is initialized as follows:

$KP(i) = 1$ if the $i$th parameter is known, $KP(i) = 0$ if the $i$th parameter is unknown, and $KP(i) = -1$ if the $i$th parameter cannot be measured or estimated. This last item, $KP(i) = -1$, is used in predictor problems to greatly decrease the number of combinations that must be examined.

### 2.3. Implementation

CRAMS is a FORTRAN program with a modular design. The relationship among the major modules of CRAMS is shown in Fig. 1. Details of the implementation of CRAMS have been given elsewhere [25a)].

The purpose of the INPUT module is to translate the input stream into an internal representation of the reaction system and its associated experimental

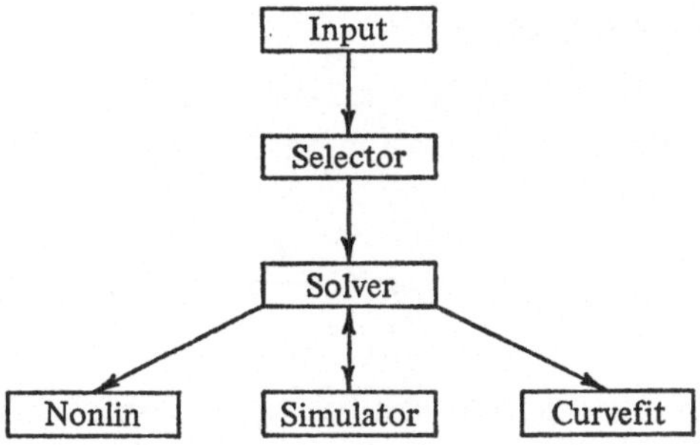


Fig. 1. Flowchart of CRAMS

data. The actual internal representation of the input stream consists of arrays that roughly correspond to the vectors and matrices defined in Section 2.2.

The SELECTOR module performs two different functions, depending on whether the problem is of the computing or predictor type. The default type of problem is the computing type. An input flag, set by the user, selects the predictor type problem. If the problem is of the computing type, SELECTOR manipulates the equations into a set of equations to be solved by the SOLVER module. Also, if there are uncomputable parameters, SELECTOR will eliminate those equations that contain them. In the present implementation of the SELECTOR module, only manipulations with the *FLUX* matrix are used to form the solvable set of equations. Manipulations with the rate expressions are not performed because no symbol manipulation capability presently exists in the system. If the problem is of the predictor type, parts of the SELECTOR module are called repeatedly, testing a great number of parameter combinations to determine what parameters become solvable. The design of the SELECTOR module is described in greater detail in Part 3.

The SOLVER module is the communications link between the three numerical analysis service modules: NONLIN, SIMULATOR, and CURVEFIT. SOLVER solves the equations that were chosen by SELECTOR by using (1) NONLIN — to initially bring the system to equilibrium, (2) SIMULATOR — to generate concentration data for certain unknown variable parameters and (3) CURVEFIT — to solve for unknown constant parameters and to test the mathematical validity of the proposed reaction model. The SOLVER module has been designed so that the three numerical analysis service modules are easily replacable as more advanced techniques are developed. The design of the SOLVER module is described in detail in Part 4. The modules NONLIN, SIMULATOR, and CURVEFIT are discussed in 4.2., 4.3., and 4.4., respectively.

The design of the INPUT module has been greatly influenced by a study of the systems mentioned in Section 1.2. The principal new contribution in this paper is the design of the SELECTOR and SOLVER modules. The numerical analysis modules are based on existing subroutines (in the case of the SIMULATOR and CURVEFIT modules) or on well known methods (in the case of the NONLIN module).

### 2.4. Illustration of New Concepts

The purpose of this section is to give a general description of the new concepts that are used in CRAMS. For illustrative purposes suppose the proposed reaction system model is given as:

$$A + B \underset{K2}{\overset{Kl}{\rightleftharpoons}} C \xrightarrow{K3} A + D$$

The coefficients of any reactant can be any real number, and any type of rate expression may be specified explicitly. However, in this illustration it is supposed that only single molecules of A, B, C, and D react. The input language for CRAMS [25b)] is used to enter information about the model and its associated data thusly:

```
$ RATE REACTIONS
   A + B = C, K1, K2;
   C = A + D, K3;
$ CONSTANTS
   K3 = 3.0;
$ INITIAL CONCENTRATIONS
   A = 1;
   B = 2;
$ DEFINE
   # A = .01 * A;
$ DATA
   (TIME, A)
   0.00 1.00
   0.05 1.54
      .
      .
      .
   0.45 2.85
$ STOP
```

The card that follows the $ RATE REACTIONS card describes a reversible reaction. The forward reaction is one unit of reactant A combining with one unit of reactant B to form one unit of reactant C. The backward reaction is one unit of C forming one unit of A and one unit of B. The forward and backward rate constants are $K1$ and $K2$ respectively.

The forward and backward rate expressions are (by default) [A]*[B] and [C] respectively. The next card describes the reaction of one unit of C reacts to form one unit of A and one unit of D. The rate constant is $K3$ and the rate expression is [C]. The card after the $ CONSTANTS card sets a rate constant, $K3$, to the value 3.0. The cards following the $ INITIAL CONCENTRATION card set the values for the initial concentrations of A and B. By default, the initial concentrations of C and D are set to zero. Following the $ DEFINE card is a card which sets the maximum tolerance of A to one percent. The cards following the $ DATA card give concentration versus time data for A. Thus, for example, at TIME = .05, the observed concentration of A is 1.54.

With $X(1)=\text{A}$, $X(2)=\text{B}$, $X(3)=\text{C}$, and $X(4)=\text{D}$, where $X(i)$ is the label of the $i$th reactant, the differential equations representing the system are:

$$\begin{aligned} CP(1) &= -K1*C(1)*C(2) + K2*C(3) + K3*C(3) && \ldots \text{A} \\ CP(2) &= -K1*C(1)*C(2) + K2*C(3) && \ldots \text{B} \\ CP(3) &= K1*C(1)*C(2) - K2*C(3) - K3*C(3) && \ldots \text{C} \\ CP(4) &= K3*C(3) && \ldots \text{D} \end{aligned}$$

Here $CP(i)$ is the rate of change in the concentration of the $i$th reactant, and $C(i)$ is the concentration of the $i$th reactant. The INPUT module creates the $FLUX$ matrix for the proposed reaction system, thus:

$$FLUX = \begin{array}{c} \\ \text{A} \\ \text{B} \\ \text{C} \\ \text{D} \end{array} \begin{array}{c} \begin{array}{ccc} K1 & K2 & K3 \end{array} \\ \left\| \begin{array}{ccc} -1 & 1 & 1 \\ -1 & 1 & 0 \\ 1 & -1 & -1 \\ 0 & 0 & 1 \end{array} \right\| \end{array}$$

The $FLUX$ matrix is a convenient way to concisely represent systems of equations and representations of reaction systems. However, the rules for manipulating the $FLUX$ matrix to formulate a solvable set of equations are complex, and they are the subject of much of the research presented in this paper. In the SELECTOR module, the $FLUX$ matrix is manipulated in such a way as to: (1) reduce the number of differential equations representing the system and (2) allow for both variable and constant parameters to be used in the computation, and (3) make the calculation on the equilibrium portion of the model considerably more efficient. The first two concepts are illustrated next with the reaction model given above. The algorithm that is used to automatically accomplish these objectives is discussed in Section 3.2.

SELECTOR first observes that rows three and four of the $FLUX$ matrix are linearly dependent on rows one and two. Thus:

$$\begin{aligned} \text{row}(3) &= -\text{row}(1) \\ \text{row}(4) &= \text{row}(1) - \text{row}(2) \end{aligned}$$

SELECTOR then performs the trivial integration of these two equations. Thus:

$$C(3) = IC(3) - C(1) + IC(1)$$
$$C(4) = IC(4) + C(1) - IC(1) - C(2) + IC(2)$$

Here $IC(i)$ is the initial concentration of the $i$th reactant. $C(j)$ is the concentration of the $j$th reactant. Thus, for this reaction model, only two differential equations must be solved instead of the original four. By substituting the first differential equation [ the equation for $CP(1)$] into the second differential equation [the equation for $CP(2)$], SELECTOR obtains:

$$CP(2) = CP(1) - K3*C(3)$$

This new differential equation for $CP(2)$ is solved by the SIMULATOR module for the parameters $C(2)$, $C(3)$, and $C(4)$, at default or specified points in time. During the course of this simulation phase, required values for $C(1)$ and $CP(1)$ are calculated using numerical interpolation and differentiation. The CURVEFIT module then solves the equation for $CP(1)$ for $K1$ and $K2$. This curve fitting part also involves a test for mathematical validity, using as its basis the user's estimate of the reliability of the data.

To illustrate the use of the NONLIN module consider the following example:

$$\mathrm{A} + \mathrm{B} \overset{EK}{\longleftrightarrow} \mathrm{C} \overset{K3}{\longrightarrow} \mathrm{A} + \mathrm{D}$$

The input cards describing this model are:

$ RATE REACTIONS
C = A + D, $K3$;

$ EQUILIBRIUM REACTIONS
A + B = C, $EK$;

$ CONSTANTS
$K3$ = 3.0;
$EK$ = 2.0;

$ INITIAL CONCENTRATIONS
A = 1;
B = 2;

$ STOP

This problem requires that concentration versus time data for A, B, C, and D be generated. The original *FLUX* matrix is essentially the same as that of the previous problem, but SELECTOR would manipulate the *FLUX* matrix in such a way as to eliminate the rate constants associated with the equilibrium reaction. In the SOLVER module, NONLIN first brings the system to equilibrium, then SIMULATOR generates concentration versus time data for A, B, C, and D.

Examples illustrating the predictor and partial solution capabilities of the system are given in Sections 3.4. and 3.3., respectively.

## 3. Design of the Selector Module

### 3.1. General Discussion

The SELECTOR module is responsible for transforming the internal representation of the reaction system into a form which can readily be solved by the SOLVER module. The equations that are represented by the original *FLUX* matrix, generated in the INPUT module, may be in an unsolvable form. For example, unknown constant parameters may appear in the same equation with as yet unsolved variable parameters. Also, if there are equilibrium assumptions made about certain reactions, the associated rate constants must be eliminated. Finally, if there are unsolvable parameters, they must be identified, and the associated equations must be eliminated. This process involves a rearrangement of the equations that represent the reaction system, using the *FLUX* matrix. Other rearrangements may be possible by examining the rate expressions, but the symbol manipulative capability that is needed to accomplish this is not yet available in CRAMS.

A secondary benefit from the manipulations that are performed to yield solvable equations is that the resulting equations may allow a more efficient calculation by the SOLVER module. For example, in the reaction system given in Section 2.4., a system of four differential equations was transformed into a system of two first order ordinary differential equations and two linear algebraic equations.

SELECTOR is also capable of predicting what parameters must be given in order to solve for some or all of the remaining unknown parameters. This involves the trial and error enumeration of a number of different possible solutions. The number of trial solutions depends on the results of previous trial solutions (see [25a]).

### 3.2. Transformation of the FLUX Matrix

An important part of SELECTOR, which is used to transform the *FLUX* matrix, is a modified version of the IBM subroutine *DMFGR* [26]. Utilizing Gaussian elimination, $DMFGR\,(A, m, n, r)$ is capable of determining the rank, $r$, and linearly independent rows of a given $m$ by $n$ matrix, $A$. Furthermore, nonbasic rows are expressed in terms of basic ones. This subroutine has been modified by us to favor column interchanges, and row interchanges are made only as a last resort. This modification was made because in certain cases it is desirable to favor certain rows as basic rows. For example, if the concentrations of certain reactants are known, forcing them to the top of the matrix may allow for the concentrations of other reactants to be directly computed using dependency information. For an $m$ by $n$ matrix $A$, *DMFGR* rearranges the rows such that the first $r$ rows of $A$ are linearly independent. The last $m$-$r$ rows are replaced by dependency information such that for $1 \leq i \leq m\text{-}r$,

$$\text{row}(r+i) = \sum_{j=1}^{r} A(r+i,j) * \text{row}(j) .$$

The algorithm that is used by the SELECTOR module to manipulate the original *FLUX* matrix is given next.

*Algorithm T* (Transformation of *FLUX* matrix). This algorithm transforms the *FLUX* matrix into a form suitable for SOLVER, if a sufficient number of parameters are given. The notation and definitions used in this algorithm are those discussed in Section 2.2. It uses only the *FLUX* matrix to rearrange the equations representing the reaction system into a set of solvable equations that can be solved by the SOLVER module. The objective is to arrange the equations into two sets: (1) those equations that will be used to solve for the unknown variable parameters by the NONLIN and SIMULATOR modules, and (2) those equations that will be used by the CURVEFIT module to solve for the unknown constant parameters. The first group of equations must not contain unknown constant parameters, and they must be completely solved before the second set of equations can be solved. Throughout the algorithm, when a change is made in the ordering of the rows or columns of the *FLUX* matrix, appropriate changes are also made in some of the other data structures. Any data structures that are not used in this algorithm can be rearranged later. For example, the *KP* vector, which contains a representation of the status of each parameter, must be immediately rearranged because it is used in subsequent steps of the algorithm. Thus, if the first two rows of *FLUX* are interchanged, the first two entries in *KP* are interchanged also.

T 1. [Initialize.]

*NKRK* = number of known rate constants

*NURK* = number of unknown rate constants

*NKEK* = number of known equilibrium constants

*NUEK* = number of unknown equilibrium constants

*IURK* = *NKRK*

*IKEK* = *IURK* + *NURK*

*IUEK* = *IKEK* + *NKEK*

*NBAD* = 0 (number of unsolvable parameters)

T 2. [Order reactions.] Order the columns of *FLUX* such that the rate reactions precede the equilibrium reactions, the rate reactions with known rate constants precede the rate reactions with unknown rate constants, and the equilibrium reactions with known equilibrium constants precede the equilibrium reactions with unknown equilibrium constants. At this point: (1) the first *NKRK* columns of *FLUX* represent rate reactions with known rate constants, (2) the next *NURK* columns of *FLUX* represent rate reactions with unknown rate constants, (3) the next *NKEK* columns of *FLUX* represent equilibrium reactions with known equilibrium constants, and (4) the last *NUEK* columns of *FLUX* represent equilibrium reactions with unknown equilibrium constants. This step prepares for operations that must be performed on parts of the matrix associated only with certain types of reactions, which are defined at the end of this algorithm.

T 3. [Eliminate Extra Known Equilibrium Constants.] Let *TFLUX* be the transpose of *FLUX*. Call *DMFGR* [*TFLUX* (*IKEK* + 1,1), *NKEK*, *NRCT*, *IRANK*]. For $IRANK < i \leq NKEK$, set $KP(NRCT + IKEK + i) = 0$. Set $NKEK = IRANK$ and adjust other pointers and counters to reflect this change. Set $TFLUX = FLUX$. Order the rows of *FLUX* such that unknown reactants precede known reactants. CALL *DMFGR* [*TFLUX* (1, *IKEK* + 1), *NRCT*, *NKEK*, *IRANK*]. Set $NEQV = NKEK$. This step is necessary to insure that in SOLVER, there are the same numbers of variables and equations that represent the equilibrium portion of the model. If a least squares technique were used to solve for the unknown equilibirum constants, this step would not be necessary. However, mathematical validation of the model would not then be possible.

T 4. [Find Normal Dependencies.] For row numbers greater than *NEQV*, order the rows of *FLUX* such that known reactants precede unknown reactants. Call *DMFGR* [*FLUX* (1,1), *NRCT*, *NRXN*, *RANK*]. Set $NNDP = NRCT - RANK$. For $1 \leq i \leq NNDP$, set $KP(RANK + i) = 0$. This step determines which reactants may be expressed as a linear combination of the other reactants. Known reactants were favored as basic rows in order that the concentrations for the greatest number of unknown reactants could be calculated.

T 5. [Find Equilibrium Dependencies.] CALL *DMFGR* [*FLUX* (1,*IKEK* + 1), *RANK*, *NEK*, *ERANK*]. Set $NEDP = RANK - ERANK$ and $NKEC = ERANK - NKEK$. For $1 \leq i \leq NKEC$, if $KC(NEQV + i) = 0$, then set $NBAD = NBAD + 1$ and $KC(NEQV + i) = -1$. For $1 \leq i \leq NEDP$ and $1 \leq j \leq NRK$, set $FLUX(ERANK + i, j) = FLUX(ERANK + i, j) -$

$$\sum_{k=1}^{ERANK} FLUX(ERANK + i, IKEK + k) * FLUX(k, j).$$

This step effectively eliminates rate constants from those reactions that are in equilibrium.

T 6. [Find Special Dependencies] CALL *DMFGR* [*FLUX* (*ERANK* + 1, *IURK* + 1), *NEDP*, *NURK*, *URANK*]. Set $NSDP = NEDP - URANK$ and $ISDP = ERANK + URANK$. For $1 \leq i \leq URANK$, if $KP(ERANK + i) = 0$, then set $NBAD = NBAD + 1$ and $KP(ERANK + i) = -1$. For $1 \leq i \leq NSDP$, set $KP(ISDP + i) = 0$. For $1 \leq i \leq NSDP$ and [$1 \leq j \leq$

*NKRK* or $IKEK < j \leq NRXN$) set $FLUX(ISDP+i,j) = FLUX(ISDP+i,j] -$

$$\sum_{k=1}^{URANK} FLUX(ISDP+i, IURK+k)*FLUX(ERANK+k,j).$$

This step eliminates unknown rate constants from equations that must be used to generate concentration data.

T 7. [End of Algorithm.] At this point, the *FLUX* matrix represents a set of equations that may be solved in two stages. The first stage generates concentration versus time data for unknown reactants and the second stage solves for the unknown constant parameters.

If there are no unsolvable parameters ($NBAD = 0$), then the first three types of equations must all be solved simultaneously for the unknown variable parameters.

*Type* (1)

The *NEQV* equations of this type have the form: $R(IKEK+j) = K(IKEK+j)$, with $1 \leq j \leq NEQV$. The derivative form is: $R'(IKEK+j) = 0$.

*Type* (2)

There are *NSDP* equations of the form:

$$CP(ISDP+i) = \sum_{j=1}^{NKRK} FLUX(ISDP+i,j)*K(j)*R(j) + \sum_{j=1}^{URANK} FLUX(ISDP+i,IURK+j)*CP(ERANK+j) + \sum_{j=1}^{ERANK} FLUX(ISDP+i,IKEK+j)*CP(j),$$

with $1 \leq i \leq NSDP$.

*Type* (3)

The *NNDP* equations have the form:

$$C(RANK+i) = \sum_{j=1}^{RANK} FLUX(RANK+i,j)*[C(j) - IC(j)],$$

with $1 \leq i \leq NNDP$. The derivative form is:

$$CP(RANK+i) = \sum_{j=1}^{RANK} FLUX(RANK+i,j)*CP(j)$$

*Type* (4)

The family of *URANK* simultaneous equations of this type are solved for the unknown rate constants only after the variable parameters have been calculated. The general form is:

$$CP(ERANK+i) = \sum_{j=1}^{NKRK} FLUX(ERANK+i,j)*K(j)*R(j) +$$

$$\sum_{j=1}^{ERANK} FLUX(ERANK+i, IKEK+j)*CP(j) +$$
$$\sum_{j=1}^{NURK} FLUX(ERANK+i, IURK+j)*$$
$$K(IURK+j)*R(IURK+j),$$

with $1 \leq i \leq URANK$.

*Type* (5)

Unknown equilibrium constants are computed from data for the variable parameters by solving the $NUEK$ equations of the form:

$$R(IUEK+j) = K(IUEK+j), \text{ with } 1 \leq j \leq NUEK.$$

It should be noted that $NBAD=0$ is a necessary but not a sufficient condition to guarantee solution. The existence of a solution is determined in SOLVER because there is presently no capability in the system to examine individual rate expressions. For example, with a single reversible reaction, if data for one reactant is given, it is usually possible to calculate the two rate constants. SELECTOR will always say that it is possible. However, if the rate expressions for the forward and backward reactions are identical, then the calculation cannot be done, but this can not be determined until the CURVEFIT module attempts the calculation.

### 3.3. Partial Solution

In the event that $NBAD > 0$ (there are some unsolvable parameters), the $FLUX$ matrix must be reduced to eliminate the unsolvable parameters. Initially, the first three types of equations are repeatedly subjected to the following test, until no further changes are made. If an unsolvable parameter appears in an equation (other than in a term with a zero coefficient) all other unknown parameters in that equation are marked as unsolvable. Then, the equations with no unsolvable parameters may be used to solve for the remaining solvable variable parameters. With the equations of Types (3) and (4), a similar algorithm is used, except that no variable parameters are marked as unsolvable.

This point is illustrated by the following model.

$$A + B \underset{K2}{\overset{K1}{\rightleftharpoons}} C \xrightarrow{K3} A + D$$
$$X \xrightarrow{K4} Y.$$

Using the input language described in the Operations Manual [25b)], the cards describing this model are:

$ RATE REACTIONS

A + B = C, $K1$, $K2$;

C = A + D, $K3$;

X = $Y$, $K4$;

.
.
.

$ STOP

In this case, if data for A is given, then only concentration versus time data for C may be generated. The equations for the other reactions are effectively removed from the system. Similarly, if data for A and $K3$ is given, then the reaction involving X and Y is eliminated and only the parameters B, C, D, $K1$, and $K2$ are computed.

The next example shows the kind of model for which SELECTOR may give misleading information about a partial solution because of its inability to examine the rate expressions.

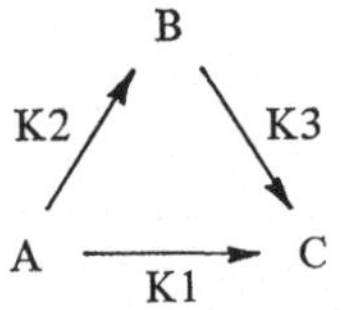


The cards describing this model are:

$ RATE REACTIONS
A = C, $K1$;
A = B, $K2$;
B = C, K3;
.
.
.
$ STOP

Assuming default rate expressions [that is, $R(1) = [A]$, $R(2) = [A]$, and $R(3) = [B]$, if data for A and B is given, SELECTOR correctly concludes that it is possible to solve for the remaining unknown parameters. However, if only data for A is given, and since the first two reactions have the same rate expression, SELECTOR incorrectly assumes that $K1$ and $K2$ can be computed. If the rate expressions were different, then the calculation would have been possible. This example emphasizes that SELECTOR only identifies those reactants that are needed to compute the remaining parameters. Thus when the system reduces to an overspecified reaction system, misleading information may be given.

### 3.4. Predictor Capabilities

At the user's request, CRAMS is capable of enumerating possible solutions by calling parts of SELECTOR repeatedly.

The input vector *KP*, which is constructed from the input data, contains the status of each parameter as follows:

(i) $-1$ means the parameter cannot be meaningfully computed or measured and it should never be considered by SELECTOR to be 'known'. The concentrations of reaction intermediates produced by very fast reactions is a typical example of this kind of parameter.

(ii) 0 means the parameter can be computed or measured but that its value is not known.

(iii) 1 means that value(s) for the parameter are known. With each call SELECTOR alters *KP* so that eventually all combinations of 0 entries are changed to 1.

If the trial *KP* is a superset of a complete solution, SELECTOR is not called.

This capability permits an experimental scientist to determine what additional data is needed by the system to complete the calculation of unknown parameters. By informing the system of parameters that have already been determined, or those parameters which can not be measured, the number of choices is considerably reduced. Thc following example illustrates this point.

$$\mathrm{A} + \mathrm{B} \underset{K2}{\overset{K1}{\rightleftharpoons}} \mathrm{C} \xrightarrow{K3} \mathrm{A} + \mathrm{D}\,.$$

The input cards describing this model are:

$ SYSTEM PARAMETERS
*MODE* = 1;
$ RATE REACTIONS
A + B = C, *K*1, *K*2;
C = A + D, *K*3;
.
.
.
$ STOP

If no data is given, SELECTOR prints, among several partial solutions, the following minimal combinations that are needed to compute the remaining unknown parameters.

| | | | |
|---|---|---|---|
| (1) | A, B | (2) | B, C |
| (3) | A, D | (4) | B, D |
| (5) | C, D | (6) | A, *K*1, *K*2 |
| (7) | C, *K*1, *K*2 | (8) | D, *K*1, *K*2 |
| (9) | A, *K*3 | (10) | B, *K*3 |
| (11) | C, *K*3 | (12) | *K*1, *K*2, *K*3 |

Suppose $K3$ had been given a positive value in the input stream as follows:

$ CONSTANTS

$K3 = 1;$

These cards indicate that $K3$ can be easily measured or is known from previous experiments. In this case only those combinations listed above that contain $K3$ would be printed. Now suppose $K3$ had been set to a negative value as follows:

$ CONSTANTS

$K3 = -1;$

These cards indicate that $K3$ is difficult or impossible to measure experimentally. In this case, only the combinations listed above that do not contain $K3$ would be listed. This facility offers a significant savings of computer time by substantially reducing the number of different combinations of the $KP$ vector that have to be generated by SELECTOR. For the above example, two additional combinations of parameters are sufficient to calculate all the remaining unknown parameters, but are not recognized by the current version of CRAMS because they depend on the rate expressions. These are:

(13) B, $K1, K2$ (14) D, $K3$

It is known that these two cases may be solved by using *ad hoc* methods that involve a detailed examination of the rate expressions. However, the current version of CRAMS cannot recognize, and, therefore, cannot solve these two cases. This problem is discussed in Section 6.2.

If $NZ$ is the number of zero entries in $KP$ then there are $2^{**}NZ$ possible configurations of $KP$. Each one of these possible configurations may be represented by a different integer number. The $i$'th configuration of $KP$ is obtained by replacing the zero entries in $KP$ by the reverse of the binary representation of the integer number $i$. A list of integer numbers representing total solutions is maintained to eliminate testing for supersets. Before $KP$ is tested by the other modules of SELECTOR, values that were set to $-1$ by the INPUT module are changed to zeros.

For example, if $KP = (-1,0,1,-1,0,1)$, there are the 2**2 or 4 possible unique configurations:

configuration 0 : $(-1,0,1\ -1,0,1)$

configuration 1 : $(-1,1,1,-1,0,1)$

configuration 2 : $(-1,0,1,-1,1,1)$

configuration 3 : $(-1,1,1,-1,1,1)$

If it were found that configuration number 1 was a total solution, for example, then configuration number 3 would not have been tested.

## 4. Design of the Solver Module

### 4.1. General Discussion

The SOLVER module is responsible for solving the equations represented by the final form of the *FLUX* matrix created by the SELECTOR module, as described in Section 3. SELECTOR determines the kind of equation that is to be solved, then SOLVER coordinates the execution of the three numerical analysis service modules: NONLIN, SIMULATOR and CURVEFIT, which are always executed in that order. The five different types of equations handled by SOLVER have been given in Section 3.2.

### 4.2. The NONLIN Module

The NONLIN module is responsible for intializing the concentration vector, $C(i)$, for $1 \leq i \leq NRCT$. Here $NRCT$ is the number of reactants. If there are no equilibrium reactions, then $C(i)$ is set to $IC(i)$, the initial concentration vector, for $1 \leq i \leq NRCT$. If equilibrium reactions do exist, then the type (2) equations (with derivatives set to zero) and the Type (1) and Type (3) equations are all solved simultaneously for the equilibrium concentrations of all reactants. Because the equilibrium equations are generally nonlinear, the Newton-Raphson iteration method [27] is used to solve these equations. Also, since there is no symbol manipulation capability in the current version of CRAMS, numerical differentiation is used to calculate the required partial derivatives. That is, the rate expressions cannot at this time be automatically differentiated by analytical methods. A three point differentiation formula is used [27]:

$$f'(x) = \frac{f(x+h) - f(x-h)}{2h}$$

The Newton-Raphson iteration method is terminated when either the absolute values of the residuals and the differences between successive iterates are less than a specified tolerance, or when the specified maximum number of iterations is exceeded.

### 4.3. The SIMULATOR Module

In the SIMULATOR module, concentration data for solvable unknown variable parameters is generated by numerically solving the equations of Types (1), (2) and (3) simultaneously. The derivative forms of the Type (1) equations are used; thus, there are $NSIM = NEQV + NSDP$ differential equations and $NNDP$ algebraic equations to be solved. $NEQV$, $NSDP$ and $NNDP$ are the numbers of Type (1), Type (2) and Type (3) equations respectively (see end of Section 3.2.). The IBM subroutine DHPCG [26] is used for this purpose.

The initial step size is computed from the initial time, the final time and the number of points for which output is requested. An error message is printed if the step size is less than a prespecified amount.

Each function evaluation by DHPCG requires a solution to the following problem: Given the value of TIME and the NSIM values of $C(i)$, calculate the NSIM values of $CP(i)$ for all $1 \leq i \leq NEQV$ and $ISDP < i \leq NSDP$. Each function evaluation follows a four step procedure that involves: (1) calculating the concentrations of the known reactants and their associated maximum tolerances, (2) calculating the derivatives of the known reactants and their associated maximum tolerances, (3) calculating the concentrations of dependent reactants, and (4) calculating the derivatives of the NSIM reactants being simulated.

The concentrations and edrivatives of the concentrations of known reactants are calculated for a normalized time scale by using Lagrangian interpolation [27]. The normalized time scale is assigned by the user [25b] or it defaults to that for the first reactant for which data was entered [25a]. First, the $NDEG$ and $NDEG - 1$ nearest points are found, where $NDEG$ is the prespecified degree of the interpolating polynomial. In the next step, if data for the $i$th reactant is unknown, its concentration is calculated from

$$C(i,t) = \sum_{j=1}^{NDEG} L(j,t)*C[i,t(j)] \, .$$

In this equation, $C(i,t)$ is the concentration of the $i$th reactant at $TIME = t$. The $L(j,t)$'s are the Lagrange coefficient polynomials. The $C[i,t(j)]$'s are given as the input data and are the concentrations of the ith reactant measured at $TIME = t(j)$. In the next step a similar formula is used to calculate $C(i,t)$ from $NDEG - 1$ nearest neighbors and then $DC(i,t)$, the maximum tolerance of $C(i,t)$, is calculated by interpolation from the input values for $DC$, $DC[i,t(j)]$. A check is made to insure that the difference between the calculated $C(i,t)$'s for the $NDEG$ and the $NDEG - 1$ points (the truncation error) is less than one percent of $DC(i,t)$, the maximum tolerance for the observed variable. Because $DC[i,t(j)]$ is normally specified by the user and $DC(i,t)$ is computed from it, this check guarantees that truncation and roundoff errors are insignificant when compared with the user supplied experimental errors and assures that the model validation can be safely postponed until the CURVEFIT module is executed.

The derivatives and their associated maximum tolerances for the known reactants are calculated by using numerical differentiation formulas.

These formulas can be derived by differentiating the interpolation formulas used for calculating $C(i,t)$. The formula for calculating $CP(i,t)$ is:

$$CP(i,t) = \sum_{j=1}^{NDEG} L'(j,t)*C[i,t(j)]$$

The truncation error is calculated by comparing this result with a calculation using one less point. The derivative of the maximum tolerance, $DCP(i,t)$, is calculated from:

$$DCP(i,t) = \sum_{j=1}^{NDEG} L'(j,t)*DC[i,t(j)]$$

A check is again made to insure that the truncation error is less than one percent of the calculated maximum tolerance.

The third step in the function evaluation by DHPCG involves the calculation of dependent concentrations. This is done by evaluating the non-derivative form of the Type (3) equations.

In the final step DHPCG calculates the NSIM derivatives for the reactants that are being simulated. Since the derivative forms of Type (1), (2) and (3) equations are all linear with respect to the $CP(i)$'s, they are solved simultaneously for the $CP(i)$'s by the Gaussian elimination method. The partial derivatives used in the evaluation of the derivative form of the Type (1) equation are calculated with the same numerical differentiation formula that is used in the NONLIN module.

Upon completion of the execution of the SIMULATOR module, all variable parameters have been calculated; *i.e.*, data for the concentrations for all known reactants has been generated to the normalized time scale.

The interface between DHPCG and the rest of the SIMULATOR module has been designed os that DHPCG can be easily replaced by better subroutines when they become available. Several methods, such as those by Gear [28] and Liniger and Odeh [29] that have been written especially for stiff differential equations, would appear to be likely candidates for replacing DHPCG. This is true because the differential equations that are obtained from chemical reactions are frequently very stiff.

### 4.4. The CURVEFIT Module

Unknown rate and equilibrium constants in Type (4) and Type (5) equations are computed by the curve-fitting module CURVEFIT. In the current version of CRAMS there is a choice of two curve-fitting methods, DLLSQ [26] or CURFIT [30]. However, new modules can be easily incorporated [25a]. DLLSQ uses the simple least-squares method to compute the unknown constants and it suffers from the unreliabilities and ambiguities common to all such methods [30a]. CURFIT performs a complete analysis of the experimental data (which includes user-supplied maximum tolerances) and tests the mathematical validity of the proposed reaction model, then computes the unknown constants and their associated maximum errors [30a]. Since the second choice is a superset of the calculations performed by DLLSQ, in the following discussion it is assumed that CURFIT is used.

Interpretation of physical data is a two step process involving: (1) the formulation of a model and (2) the formulation of equations describing the model. Certain physical restrictions, like the conservation of mass etc., are an intrinsic part of the formulation of a proposed model. The meanings attached to the formulation of the model itself must be taken into consideration when formulating the equations that describe the model.

The verification of the proposed model is also a two step process involving: (1) mathematical validation (the process of verifying that the equations are valid algebraic descriptions of the data that is used to test the model) and (2) physical validation (the process of verifying that the mathematically validate equations also meet the physical conditions imposed by the model). It is not generally realized that the validity of a model is established only if both the mathematical and physical validities are established. The deplorable and common technique of using physical validity alone to establish a model's validity has frequently yielded erroneous conclusions [30a)].

In the CURFIT module the mathematical validity of those equations that represent those parts of the model that contain unknown constant parameters is tested in terms of the user's estimate of the reliability of his data. However, it should be noted that mathematical validity does not, by itself, establish the worth of the model and that CURFIT does not test for physical validity of the model. This can, in general, only be performed by one who is an expert in the physical meaning of the model.

The solution of all the Type (4) equations for a model is accomplished by a single use for the CURFIT module. Each Type (4) equation is multiplied by its left hand side; then they are all added together. Data at each required time may be generated from this summed equation and the maximum tolerances for the dependent variable are calculated with a similar equation.

Because the left hand side of Type (4) equations are simply derivatives of concentrations of reactants, the maximum tolerances of the dependent variable of the equation that is used to solve for the rate constants is simply the sum of squares of the maximum tolerances of the derivatives. Thus, even if terms on the right hand side are known, their errors are considered to be zero. In other words, the maximum tolerance associated with a dependent variable of an individual equation is always calculated in the same way, regardless of how many of the rate constants are known. This method of calculation effectively assumes that any data generated by the SIMULATOR module is considered to be exact. It should be noted that this assumption leads to a more restricted solution than that which would be obtained if maximum errors in all the known terms (that is, those with constant parameters) had been considered.

Each Type (5) equation is solved by one separate use of the CURFIT module. In the present version of CRAMS, the maximum tolerance for each dependent variable is arbitrarily set to one percent of its value. This is done because some of the variable parameters in the rate expressions, which are the dependent variables, may have been generated by one of the other modules and at this time such cases cannot be identified. However, it is planned to implement a symbol manipulating capability that can identify such cases. When this capability is implemented all the options allowed for specifying maximum tolerances [30b)] will be allowed.

The goodness of fit criteria that are used in CURFIT are as follows: (1) If the data are not described by selected equation, CURFIT returns the conclusion that the data is described by two or more equations (of the selected form) with overlapping domains. In this case the domains, parameters, and the associated maximum errors for each equation are given. (2) If the data are described by the selected equation, CURFIT computes its parameters and their associated maximum errors. Bad data points are automatically rejected. Thus the number of equations returned by CURFIT determines whether or not the data is described by the proposed reaction model. In those cases where the model is not described by all the data, the information returned by CURFIT can be used to specify what subset(s) of the data fits the reaction model.

## 5. Illustration of the Use of CRAMS

### 5.1. Description of the Problem

Recent work has been directed toward exploring the behavior of the enzyme aconitase [31–33]. Presently, CRAMS is being used to obtain an insight into the mechanisms of reactions involving this enzyme [34]. In this part, the problem is first stated in general terms, then a step by step description of the use of CRAMS is given.

It is known that the enzyme aconitase catalyzes: (1) the dehydration of both citrate and isocitrate to form cis-aconitate, (2) the reverse reactions, and (3) the interconversion of citrate and isocitrate thusly:

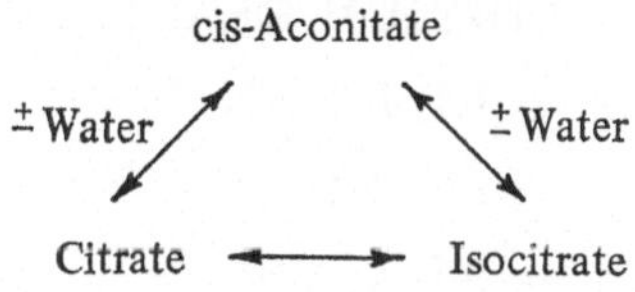


Experiments indicate that there may be two forms of the enzyme. Two possible mechanisms could explain this theory. The problem is to: (1) determine which, if any, of the two proposed mechanisms explains the theory, and (2) determine the ratio of the two forms of the enzyme.

The following abreviations are used throughout the remainder of the discussion:

E – enzyme (aconitase)
X – alternate form of aconitase
H – hydrogen ion (proton)
C – citrate
I – isocitrate
A – cis-aconitate

Combinations of these abreviations, such as EC, are intermediate complexes involving the reactants associated with the individual symbols.

The two proposed reaction mechanisms are:

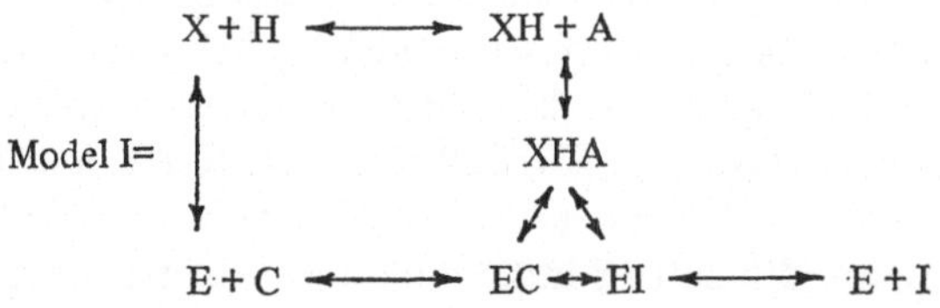


and:

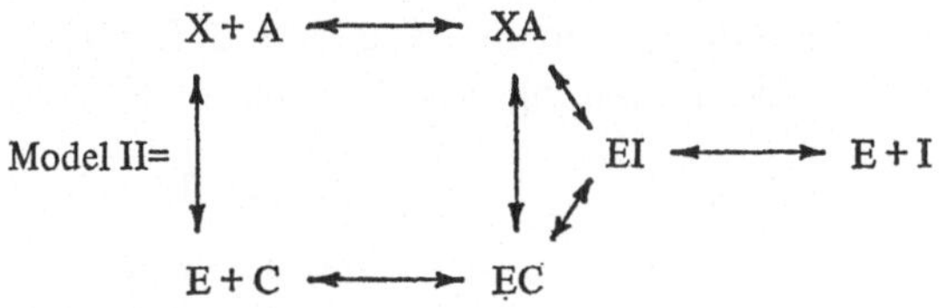

### 5.2. Solution of the Problem

First, both models were run with CRAMS with MODE = 1 (predictor problem), assuming all reactions in equilibrium. Then, both models were again run with MODE = 1, but with both rate and equilibrium reactions representing the actual time course behavior of the models.

(1) Model I, all equilibrium

```
$ SYSTEM PARAMETERS
  MODE = 1;
$ EQUILIBRIUM REACTIONS
  XHA = EI, K1;
  EI  = EC, K2;
  EC  = XHA, K3;
  EI  = E + I, KI;
  XHA = XH + A, KHA;
  EC  = E + C, KC;
  XH  = X + H, KH;
  E   = X, KE;
$ CONSTANTS
  K1  =  1;    K2  =  1;    K3 =  1;
  KI  =  1;    KHA =  1;    KC =  1;
  XHA = -1;    EI  = -1;    EC = -1;
  XH  = -1;    E   = -1;    X  = -1;
$ STOP
```

The card after the \$ SYSTEM PARAMETERS card indicates that the problem is a predictor type problem. The cards following the \$ EQUILIBRIUM REACTIONS card describe the proposed model to the system. The cards succeeding the \$ CONSTANTS card describe which parameters are known and which parameters can not be determined. $K1$, $K2$, $K3$, $KI$, $KHA$, and $KC$ are all known from other experiments. XHA, EI, EC, and XH are intermediate complexes and their concentrations are difficult or impossible to measure. Since it is conjectured that there are two forms of the enzyme data for E and X cannot be given either.

Given the model and the parameter specifications, CRAMS prints the following combinations that, in addition to $K1$, $K2$, $K3$, $KI$, $KHA$, and $KC$, will be needed to solve for all of the remaining unknown parameters:

| | |
|---|---|
| (1) I, H | (2) A, H |
| (3) C, H | (4) I, $KH$ |

| | |
|---|---|
| (5) A, $KH$ | (6) C, $KH$ |
| (7) H, $KH$ | (8) I, $KE$ |
| (9) A, $KE$ | (10) C, $KE$ |
| (11) H, $KE$ | (12) $KH$, $KE$ |

Since the concentration of C, A, I, and H are all relatively easy to measure, any of the first three combinations appears a reasonably candidate for computing information needed in next step [see (3) below]. A check on the data can be made by running all three combinations and comparing the results.

(2) Model II, all equilibrium

```
$ SYSTEM PARAMETERS
  MODE = 1
$ EQUILIBRIUM REACTIONS
  XA = EA,    K1;
  EI = EC,    K2;
  EC = XA,    K3;
  EI = E + I, KI;
  XA = X + A, KA;
  EC = E + C, KC;
  E  = X,     KE;
$ CONSTANTS
  K1 =  1;   K2 =  1;   K3 =  1;
  KI =  1;   KA =  1;   KC =  1;
  XA = -1;   EI = -1;   EC = -1;
  E  = -1;   X  = -1;
$ STOP
```

The cards for this model have similar meaning, to those for Model I. In addition to the given parameters (that is $K1$, $K2$, $K3$, $KI$, $KA$, and $KC$), the following combinations of parameters will be needed to solve for all of the remaining unknown parameters:

| | |
|---|---|
| (1) A | (2) C |
| (3) I | (4) $KA$ |

Since the concentrations of C, A, and I are all relatively easy to measure, any of the first three combinations appears reasonable to compute information needed in the next step [see (4) below]. A check on the data can be made by running all three possibilities.

(3) Model I, time course behavior

```
$ SYSTEM PARAMETERS
  MODE = 1;
$ RATE REACTIONS
  XHA  = EI, K1F, K1B;
  EI   = EC, K2F, K2B;
  EC   = XHA, K3F, K3B;
  XH   = X + H, KHF, KHB;
  E    = X, KEF, KEB;
$ EQUILIBRIUM REACTIONS
  EI   = E + I, KI;
  XHA  = XH + A, KHA;
  EC   = E + C, KC;
$ CONSTANTS
  K1F  =  1;   K1B  =  1;   K2F  =  1;
  K2B  =  1;   K3F  =  1;   K3B  =  1;
  KI   =  1;   KHA  =  1;   KC   =  1;
  XHA  = -1;   EI   = -1;   EC   = -1;
  XH   = -1;   E    = -1;   X    = -1;
$ STOP
```

The following combinations of parameters are sufficient to solve for the remaining unknown parameters:

(1) H, *KEF*, *KEB* (2) *KHF*, *KHB*, *KEF*, *KEB*

Unfortunately, neither of these two combinations ocntains only parameters that are easily measured or well known. However, with the first combination, H is easily measured and the equilibrium constant, *KEF* and *KEB* can be calculated from run (1). Estimates must be made for one of the rate constants, and the model must be run several times.

(4) Model II, time course behavior

```
$ SYSTEM PARAMETERS
  MODE = 1;
$ RATE REACTIONS
  XA = EI, K1F, K1B;
```

EI = EC, *K2F*, *K2B*;

EC = XA, *K3F*, *K3B*;

E = X, *KEF*, *KEB*;

$ EQUILIBRIUM REACTIONS

EI = E + I, *KI*;

XA = X + A, *KA*;

EC = E + C, *KC*;

$ CONSTANTS

| | | | | | | | | |
|---|---|---|---|---|---|---|---|---|
| *K1F* | = | 1; | *K1B* | = | 1; | *K2F* | | 1; |
| *K2B* | = | 1; | *K3F* | = | 1; | *K3B* | = | 1; |
| *KI* | = | 1; | *KA* | = | 1; | *KC* | = | 1; |
| EI | = | −1; | EC | = | −1; | XA | = | −1; |
| E | = | −1; | X | = | −1; | | | |

$ STOP

In this case, only one combination of parameters is possible:

(1) *KEF*, *KEB*

As previously shown, the equilibrium constant can be computed from run (2), leaving only one unknown rate constant. The single unknown constant can be varied to see if the mathematical validity is established.

All of the cases listed by CRAMS in each of the four runs have been tested using simulated data. CRAMS is currently being used to compare the two proposed mechanisms by using actual experimental data as suggested by the above results.

The Benkovic group has utilized CRAMS to generate the concentration of enzyme-ligand species as a function of enzyme and ligand concentrations. The enzymes in question are multi-subunit enzymes that bind more than one ligand. For example, in the case of hexose diphosphatase, the enzyme contains four subunits, each capable of binding a sugar phosphate and a metal ion. In the total solution, then, the concentrations of approximately twenty-eight species are involved. The unique ability of CRAMS to solve a problem of these dimensions has proven invaluable. Readers further interested in these applications are directed to recently published papers [36, 37].

## 6. Conclusions

### 6.1. Importance of the Study

In this paper it has been demonstrated that the CRAMS system can directly aid the physical scientist to design experiments for testing *any* proposed reaction mechanism. In CRAMS two kinds of parameters, variable (concentration-time data) and constant (rate/equilibrium constants), are recognized. CRAMS can be used to determine what parameters must be measured to compute *any* subset of the unknown parameters (Predictor Problem) or, given *any* subset of parameters it will compute the maximum number of unknown parameters possible (Computing Problem). In other words CRAMS may be used to study *any* fragment or combination of fragments of a proposed reaction system. Moreover, the following are unique features of CRAMS:

(i) No assumptions are made concerning the relative sizes of any variable (*e.g.*, concentration) or constant (*e.g.*, specific rate or equilibrium constant) parameters. The accuracy of the computations is determined solely by the computational reliability of the service modules: NONLIN, CURVEFIT and SIMULATOR, and they are easily replaced by routines that use superior methods, when they are available.

(ii) Data for variable parameters can be collected in entirely seperate, unconnected experiments without any prior consideration of a fixed time scale. This means that data for different reactants from different sources can be combined without any change in computational accuracy. However, it is necessary that the initial conditions in each separate experiment are the same.

(iii) There are no restrictions on the molecularity of any reaction or the concentrations for any reactant or the value of any constant variable (*i.e.* specific rate/equilibrium constant). Moreover, because each individual reaction is separately entered, the user can, in effect, specify specific rate constants that are any continuous function of any variable parameter (like concentrations, light intensity, gravitational field strength, etc.).

(iv) There is no restriction on the combinations of rate and equilibrium reactions that can be used to describe a reaction model. CRAMS will automatically solve equilibrium reactions only, or rate reactions only or any combination systems.

(v) There are no restrictions on the kind of functions that may be used for "specific rate constants". This means that a "specific rate constant" can in fact be a variable parameter whose values are determined by the concentration(s) of reactants and such variables as light intensity, strengths of magnetic or gravitational fields, etc.

(vi) There are no restrictions on the homogeneity of the reactions in a reaction system. This means that the most general form of the classic diaphragm problem, in which two or more separate homogeneous reaction systems themselves interact at interfacial boundaries (like a diaphragm or a gas/liquid surface) are easily handled by CRAMS. This kind of problem will be explicitly discussed in a subsequent paper.

The approach to modeling and the computer techniques used in CRAMS should be of value in designing automatic, general systems that solve classes of problems. In fact, the FRANS system, which is now being designed/implemented in our laboratories, will use much of the same philosophy and many of the same techniques. FRANS is being designed to automatically solve for unknown parameters in systems of mathematical equations of any type. Like CRAMS, the FRANS system is to have both Predicting and Computing Problem solving capabilities.

## 6.2. Implication for Future Research

Two general areas for future research exist: (1) possible improvements to the actual implementation of CRAMS and (2) possible extensions to the types of calculations that are possible with the system.

Changes in the actual implementation of the system should be directed toward making computations more efficient or increasing the amount of storage available to data structures. As was emphasized throughout the paper, the numerical analysis service modules associated with the SOLVER module can be easily replaced as more efficient techniques are developed. A possible savings in space can be obtained by using sparse matrix techniques [35] on the *FLUX* matrix. These techniques are especially attractive when it is realized that columns of the *FLUX* matrix usually have at most four or five non-zero entries. A feature whereby the size of the main storage array is automatically set, depending only on the size of the region into which the program was loaded is also desirable. In this regard the recent development of the compilerless, machine and configuration independent *P*ortable *FORTRAN*, PFORTRAN [38] is of paramount significance. It is planned to re-implement CRAMS in PFORTRAN.

Extensions to the system in terms of the types of calculations allowed are also possible. The allowable characteristics of the rate expressions could be extended to allow unknown nonlinear constants, discontinuities, and integral and differential expressions, thereby increasing the number of models that could be analyzed by the system. Also, the system could be extended to generate equations directly from statements specifically referring to rate law assumptions.

The model validation procedures in the system could also be extended. For example, in some cases it would be desirable to have a feature which automatically compares measured data with computer generated data, serving as an additional check on the mathematical validity of the proposed model. The system could also be extended to study the effects of inherent errors in the constant parameters, possibly by executing parts of SIMULATOR several times for extreme values of each constant parameter.

Another possible area of research involves exploring the feasibility of incorporating symbol manipulation techniques into the system. By symbolically differentiating certain rate expressions only once, the SIMULATOR module would become more efficient because numerical differentiation at each step would be avoided. Also, the solutions recognized by SELECTOR could be analyzed to insure that a sufficient number of parameters have been given before the execution of the CURVEFIT module. Finally, symbol manipulation techniques would allow a greater number of solutions to be recognized by the SELECTOR module. For example, recall the following model from Section 3.4.:

$$\mathrm{A} + \mathrm{B} \underset{K3}{\overset{K1}{\rightleftharpoons}} \mathrm{C} \xrightarrow{K2} \mathrm{A} + \mathrm{D}$$

There are two possible total solutions that the present implementation of CRAMS does not recognize. Here the ad hoc method of solution is given for one of these cases, that is, when data for $K3$ and D is given. With $X(1) = \mathrm{A}$, $X(2) = \mathrm{B}$,

$X(3) = \mathrm{C}$, and $X(4) = \mathrm{D}$, where $X(i)$ is the label of the $i$th reactant, the differential equations representing the system are:

$$\begin{aligned} CP(1) &= -K1*C(1)*C(2) + K2*C(3) + K3*C(3) \\ CP(2) &= -K1*C(1)*C(2) + K2*C(3) \\ CP(3) &= \phantom{-}K1*C(1)*C(2) - K2*C(3) - K3*C(3) \\ CP(4) &= \phantom{-K1*C(1)*C(2) - K2*C(3) - {}} K3*C(3) \end{aligned}$$

$CP(i)$ is the rate of change in the concentration of the $i$th reactant, and $C(i)$ is the concentration of the $i$th reactant. It is seen that the equations for $CP(1)$ and $CP(2)$ are linearly dependent on the equations for $CP(3)$ and $CP(4)$. Thus:

$$\begin{aligned} CP(1) &= -CP(3) \\ CP(2) &= -CP(3) + CP(4) \end{aligned}$$

Integration yields:

$$\begin{aligned} C(1) &= IC(1) - C(3) + IC(3) \\ C(2) &= IC(2) - C(3) + IC(3) + C(4) - IC(4) \end{aligned}$$

$IC(i)$ is the initial concentration of the $i$th reactant. Since data for $C(4)$ is given, data for $CP(4)$ can be generated using numerical differentiation. The following three equations could be solved simultaneously for $C(1)$, $C(2)$, and $C(3)$ for any point in time:

$$\begin{aligned} C(1) &= IC(1) - C(3) + IC(3) \\ C(2) &= IC(2) - C(3) + IC(3) + C(4) - IC(4) \\ CP(4) &= K3*C(3) \end{aligned}$$

After concentration versus time data for A, B, and C have been generated, the CURVEFIT module may be used to solve the equation for $CP(3)$ for the rate constants $K1$ and $K2$.

Although the solution to this particular problem is fairly straightforward, at this time, the extension of this method to the general model appears to be very difficult.

*Acknowledgements*. The authors wish to acknowledge Professors S. J. Benkovic and J. J. Villafranca and the members of their research groups for their participation in the testing of the computer program. The authors are especially indebted to John A. Lucas, II for his help in preparing this paper for publication.

## 7. References


1) Garfinkel, D., Sack, R.: Digital Computer Simulation of an Ecological System, Based on a Modified Mass Action Law. Ecology *45*, 502 (1964).

2) Garfinkel, D., Garfinkel, L., Pring, M., Green, S. B., Chance, B.: Computer Applications to Biochemical Kinetics. Ann. Rev. Biochem. *39*, 473 (1970).

3) Detar, D. F., Detar, C. E.: In: Computer Programs for Chemistry, *2* (Detar, D. F., Streitweiser, A., Weiberg, K. B., Eds.). New York: W. A. Benjamin 1967.

4) Garfinkel, D.: A Machine-Independent Language for the Simulation of Complex Chemical and Biochemical Systems. Computers Biomed. Res. *2*, 31 (1968).

5) Chance, B. M., Shepard, E. P.: Automatic Techniques in Enzyme Simulation. Computers Biomed. Res. *2* 321 (1969).

6) Rhoads, D. G., Achs, M. J., Peterson, L., Garfinkel, D.: A Method of Calculating Time-Course Behavior of Multi-Enzyme Systems from the Enzymatic Rate Equation. Computers Biomed. Res. *2*, 45 (1968).

7) Green, S. B., Garfinkel, D.: Simulation of Enzyme Systems Using a Matrix Representation. Computers Biomed. Res. *3*, 166 (1970).

8) Groner, G. F., Clark, R. L., Berman, R. A., Deland, E. C.: BIOMOD — An Interactive Computer Graphics System for Modeling. AFIPS Conf. Proc. (1971 FJCC) *39*, 369 (1971).

9) Curtis, A. R., Chance, B. M.: Numerical Methods for Simulation and Optimization. FEBS VIIth Meeting Proc. *25*, 39 (1972).

10) Chandler, J. P., Hill, D. E., Spivey, H. O.: A Program for Efficient Integration of Rate Equations and Least Sqares Fitting of Chemical Reaction Data. Computers Biomed. Res. *5*, 515 (1972).

11) deMaine, P. A. D., Seawright, R. D.: Digital Computer Programs for Physical Chemistry, Vol. II. New York: MacMillan Co. 1965.

12) Berman, M., Shahn, E., Weiss, M. F.: The Routine Fitting of Kinetic Data to Models: A Mathematical Formalism for Digital Computers. Biophys. J. *2*, 275 (1962).

13) Cleland, W. W.: Computer Programmes for Processing Enzyme Kinetic Data. Nature *198*, 463 (1963).

14) Cleland, W. W.: The Statistical Analysis of Enzyme Kinetic Data. Advan. Enzymol. *29*, 1 (1967).

15) Pring, M.: The Simulation and Analysis by Digital Computer of Biochemical Systems in Terms of Kinetic Models I. The Choice of Integration Method. J. Theoret. Biol. *17*, 421 (1967).

16) Pring, M.: The Simulation and Analysis by Digital Computer of Biochemical Systems in Terms of Kinetic Models II. Curve-Fitting Procedures. J. Theoret. Biol. *17*, 430 (1967).

17) Pring, M.: The Simulation and Analysis by Digital Computer of Biochemical Systems in Terms of Kinetic Models III. Generator Programming. J. Theoret. Biol. *17*, 436 (1967).

18) Swann, W. H.: A Survey of Nonlinear Optimization Techniques. FEBS Letters *2*, S39 (1969).

19) Kowalik, J., Morrison, J. F.: Analysis of Kinetic Data for Allosteric Enzyme Reactions as a Nonlinear Regression Problem. Mathemat. Biosci. *2*, 57 (1968).

20) Arihood, S. A., Trowbridge, C. G.: Model Selection and Parameter Evaluation by Nonlinear Regression, with an Application to Chymotrypsin Rate Data. Arch. Biochem. Biophys. *141*, 131 (1970).

21) Atkins, G. L.: Some Applications of a Digital Computer to Estimate Biological Parameters by Nonlinear Regression Analysis. Biochem. Biophys. Acta *252*, 421 (1971).

22) Sillen, L. G., Warnqvist, B.: Equilibrium Constants and Model Testing from Spectrophotometric Data, Using LETAGROP. Acta Chem. Scand. *22*, 3032 (1968).

23) DeLand, E. C.: Chemist — The Rand Chemical Equilibrium Program. Memo RM-5404-PR (Rand Corp. Santa Monica, California, 1967).

24) Bos, M., Meershoek, H. Q. J.: Computer Program for the Calculation of Equilibrium Constants in Complex Systems. Anal. Chim. Acta *61*, 185 (1972).

25) a) Butler, R. S.: The Design and Implementation of a Chemical Reaction Analysis and Modeling System. Ph. D. Thesis, The Pennsylvania State University, March, 1974;

b) Butler, R. S.: Operations Manual for CRAMS. Report Number 13, Computer Science Department. The Pennsylvania State University, March, 1974.

26) IBM Corporation: System/360 Scientific Subroutine Package. GH20-0205-4 (1968).

27) Morsund, D. G., Duris, C. S.: Elementary Theory and Application of Numerical Analysis. New York: McGraw-Hill 1967.

28) Gear, C. W.: The Automatic Integration of Ordinary Differential Equations. Comm. A.C.M. *14*, 176 (1971).

29) Liniger, W., Odeh, F.: A-Stable Accurate Averaging Multi-step Methods for Stiff Differential Equations. IBM J. Res. Develop. *16*, 335 (1972).

30) a) deMaine, P. A. D., Springer, G. K.: A Non-Statistical Program for Automatic Curve-Fitting to Linear and Non-Linear Equations. Management Informatics *3*, 233 (1974).
b) Springer, G. K., deMaine, P. A. D.: Operations Manual for the CURFIT Program. Report No. 1, Computer Science Department, The Pennsylvania State University (1973).

31) Villafranca, J. J., Mildvan, A. S.: The Mechanism of Aconitase Action, I. Preparation, Physical Properties of the Enzyme, and Activation by Iron (II). J. Biol. Chem. *246*, 772 (1971).

32) Villafranca, J. J., Mildvan, A. S.: The Mechanism of Aconitase Action, II. Magnetic Resonance Studies of the Complexes of Enzyme, Manganese (II), Iron (II) and Substrates. J. Biol. Chem. *246*, 5791 (1971).

33) Villafranca, J. J., Mildvan, A. S.: The Mechanism of Aconitase Action ,III. Detection and Properties of Enzyme — Metal — Substate and Enzyme — Metal — Inhibitor Bridge Complexes with Manganese (II) and Iron (II). J. Biol. Chem. *247*, 3454 (1972).

34) Villafranca, J. J.: The Mechanism of Aconitase Action, IV. Inhibition of Aconitase by Tricarboxylic Acids (in preparation).

35) Curtis, A. R., Reed, J. K.: The Solution of Large Sparse Unsymmetric Systems of Linear Equations. J. Inst. Math. Appl. *8*, 344 (1971).

36) Fishbein, R., Benkovic, P. A., Schray, K. J., Siewers, I. J., Steffens, J. J., Benkovic, S. J.: The Anomeric Specificity of Phosphofructokinase from Rabbit Muscle. J. Biol. Chem. *249*, 6047 (1974).

37) Libby, C.: The Mechanism of Action of Hexosediphosphatase. Ph. D. Thesis, The Pennsylvania State University, 1974.

38) a) Whitten, D. E., deMaine, P. A. D.: A Machine and Configuration Independent FORTRAN: Portable FORTRAN (PFORTRAN). IEEE Trans. Software Eng. *1*, III (1975).
b) Whitten, D. E., deMaine, P. A. D.: Operations and Logic Manual for Portable FORTRAN (PFORTRAN). Report No. 1 of a Series: Global Management Systems (1974), Computer Science Department, The Pennsylvania State University, University Park, Pennsylvania. 16802.

39) Ugi, I.: The Potential of Four Component Condensations for Peptide Syntheses — A Study in Isonitrile and Ferrocene Chemistry as well as Stereochemistry and Logics of Syntheses. Intra-Sci. Chem. Rep. *5*, 229 (1971).

Received December 4, 1975

# IR Fourier Transform Spectroscopy

**Prof. Dr. Reinhart Geick**

Physikalisches Institut der Universität, Würzburg

## Contents

## 1. Introduction

In the last two decades a "new" method of spectroscopy, namely Fourier transform spectroscopy, has been applied, not only by physicists but also inreasingly by chemists, for optical investigations, especially in the far-infrared [1–12].

Spectroscopy is in general terms the science that deals with the interaction of electromagnetic radiation with matter; in particular, it can be said to be the investigation of the optical properties, *i.e.* the transmission or reflection, of a sample within a certain spectral range. These properties are studied as a function of the wavelength or frequency of the incident electromagnetic radiation. The spectral range mainly under consideration here is the infrared.

Usually, it is divided into the middle- and near-infrared and into the far-infrared. The limits of these ranges may be defined as follows [13,14]:

| | middle- and near-infrared | far-infrared |
|---|---|---|
| wavelength $\lambda$: | 700 nm – 25 $\mu$m | 25 $\mu$m – 1 mm |
| frequency $\nu$: | $1.2 \cdot 10^{13}$ Hz – $4.3 \cdot 10^{14}$Hz | $3 \cdot 10^{11}$Hz – $1.2 \cdot 10^{13}$Hz |
| wave number $\tilde{\nu}$: | 400 cm$^{-1}$ – 14300 cm$^{-1}$ | 10 cm$^{-1}$ – 400 cm$^{-1}$ |
| quantum energy $h\nu$: | 49.7 meV – 1.77 eV | 1.24 meV – 49.7 meV |

With some optional extensions, commercial Fourier spectrometers generally operate in the range 10 – 10000 cm$^{-1}$.

In spectroscopic investigations, the intensity transmitted or reflected by the sample is compared with the intensity of the incident light at a given wavelength or frequency; thus, with a spectrometer, intensities and wavelengths are measured simultaneously. Intensities are determined by means of an infrared detector. Clearly, the performance of the detector is just as important to the spectroscopist, as that of the source, but these and other problems of infrared technique are beyond the scope of this introduction and the reader is referred to the literature [2,3].

In comparing spectroscopic methods and explaining their principles, we are concerned mainly with the problem of how to separate electromagnetic radiation into its spectral elements and how to determine the wavelength or frequency. Various methods are available. With a grating spectrometer, for example, the separation into spectral elements is effected by the diffraction of the grating. If the grating is turned round, the intensity is scanned as a function of frequency. A prism spectrometer or a Fabry-Perot works in a similar way. The problem of separation into spectral elements becomes trivial if a tunable source emitting monochromatic radiation is used. Grating or prism spectrometers are typically used for spectroscopy in the visible and near-infrared and monochromatic sources for microwave spectroscopy [29].

With the method of Fourier transform spectroscopy, the light is not separated into spectral elements. The scientist uses a two-beam interferometer and studies the interference or correlation properties of the light as a function of path difference; the results of this study are then converted mathematically to the spectrum on a computer. This conversion is a Fourier transform, which is why the method is called Fourier transform spectroscopy. All this is explained in much more detail below. The point being made here is that this method employs mathematics, computers, and electronic data processing, all perhaps strange, new tools for the spectroscopist, especially if he is a chemist. It is important to realize that the mathematical formalism of Fourier transforms is not only a necessary tool in Fourier transform spectroscopy, it also plays an important role in all fields of optics as well as in electrotechnics and in acoustics [15]. This involvement with Fourier transforms will possibly reveal some new aspects of conventional spectroscopy and optics.

It is admitted that spectroscopy in the far-infrared suffers from the lack of more powerful sources and more sensitive detectors [29]. Here Fourier transform spectroscopy has some advantages over conventional spectroscopy, *e.g.* with a grating instrument, and will probably be the most used method until coherent tunable laser sources take over.

For this reason, the reviewer proposes to introduce the reader to Fourier transform spectroscopy in the hope that he will make use of it. The basic physical principles of spectroscopy and the theory and practice of Fourier transform spectroscopy are described. Its advantages and disadvantages are discussed relative to spectroscopic problems and always with reference to the grating spectrometer as representing conventional spectroscopy.

Perhaps it is worth mentioning that the "new" method of Fourier transform spectroscopy is nearly a century old and that Michelson and Rubens already applied it in principle [16–19]. The method depends on a computer to perform the Fourier transformation and so has come into more frequent use with the advent of electronic computers [20–28]. The operation of a modern commercial Fourier spectrometer is no more complicated than that of any other spectrometer [30,31]. The spectroscopist needs no highly specialized training in programming and using the computer for the Fourier transformation. The programming has usually been done by the manufacturer so that the instrument works more or less automatically. Thus, the Fourier transformation is very much like a simple "push-button" operation. Nevertheless, it is useful to understand the fundamentals and to be able to choose the correct values for the input parameters, for even a well-programmed computer has to be fed with data and given some orders.

Obviously, the subject of Fourier transform spectroscopy cannot be treated rigorously without mentioning the mathematical aspects. Some mathematical formulae are therefore unavoidable in this introduction, but the meaning of the mathematical "shorthand notation" is always explained in the text or by means of a figure.

## 2. Fundamentals of Spectroscopy

It is proposed to recapitulate the basic physical and optical principles of spectroscopy in this review. For the comparison of different methods, we concentrate on the determination of wavelength as an essential part of spectroscopy. We also comment on the power of resolution of the various instruments and the instrument line-shape function.

### 2.1 Separation into Spectral Elements

Let us start with the determination of wavelength by means of a spectrometer. To keep the argument as simple as possible, let the source be a laser emitting monochromatic radiation of negligible linewidth, *e.g.* an HCN laser ($\lambda_0 = 337 \mu m$ or $\tilde{\nu}_0 = 29{,}67 \ cm^{-1}$). With a diffraction grating, several orders of diffraction are observed in the focal plane of mirror M3 (Fig. 1). If the grating constant $g$ and the

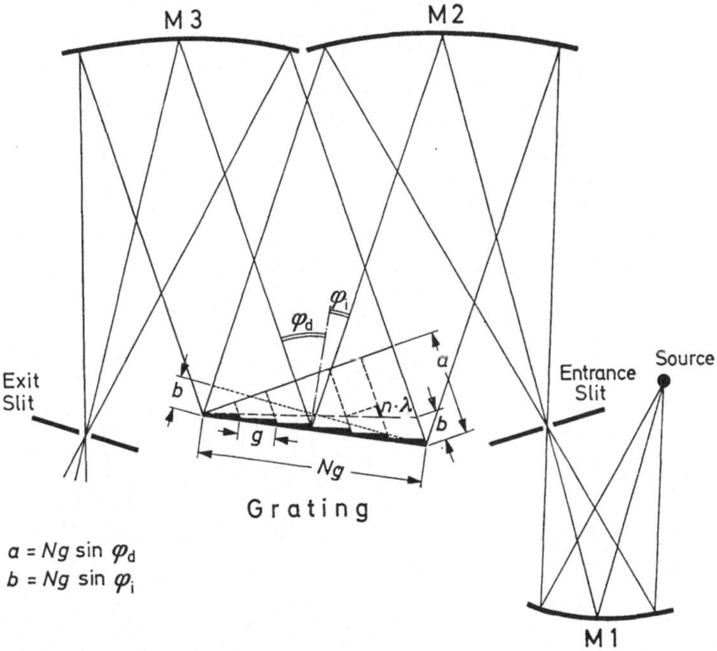


Fig. 1. Illustration of the basic formula for the diffraction grating $n\lambda = g\,(\sin \varphi_d - \sin \varphi_i)$

angle of incidence $\varphi_i$ are known, the wavelength $\lambda_0$ of the laser radiation can be determined from the diffraction angle $\varphi_d$ of the $n$-th order, according to the condition for constructive interference [32,33]:

$$n\lambda_0 = g\,(\sin \varphi_d - \sin \varphi_i) \qquad n = 0, \pm 1, \pm 2, \ldots \qquad (2.1)$$

If the light emitted by the source consists of two or more narrow lines, two or more interpenetrating sets of diffraction patterns will be obtained. From these, the wavelengths or frequencies of the lines can be deduced.

We can also determine wavelength with a Michelson interferometer (Fig. 2), where part of the light is reflected to the fixed mirror by the beam splitter and part is transmitted to the movable mirror. At the detector these two parts interfere with each other, and the interferences are governed by the position of the movable mirror or, in other words, by the difference in the optical paths of the two partial beams. Here too we consider a laser source emitting a single narrow line of wavelength $\lambda_0$: there will again be constructive interference if the path difference is an integral multiple of the wavelength [32]

$$s = n\lambda_0 \qquad n = 0, \pm 1, \pm 2, \ldots \qquad (2.2)$$

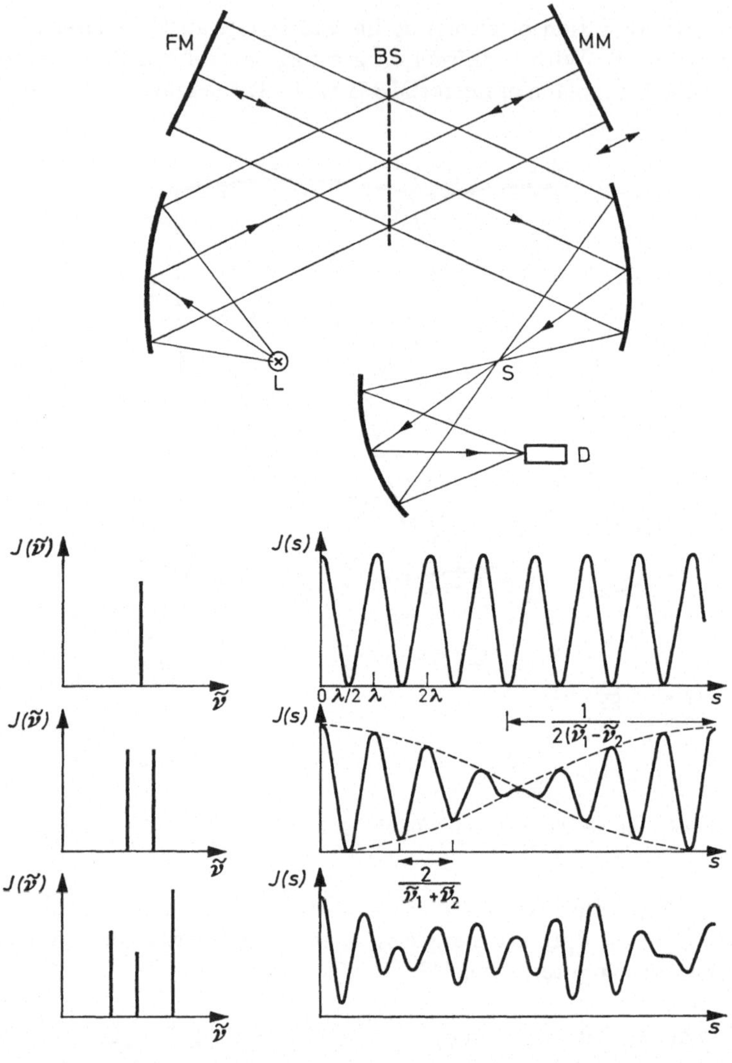


Fig. 2. Upper part: Michelson interferometer (L = source, BS = beamsplitter, FM = fixed mirror, MM = movable mirror, S = sample focus, and D = detector). Lower part: Spectra $I(\tilde{\nu})$ and interferograms $I(s)$ for one, two and three narrow laser lines

In practice, the intensity $I$ at the detector will be measured as a function of the path difference $s$, *i.e.* the interferogram $I(s)$ (see Fig. 2). The curve $I(s)$ exhibits maxima for $s = n\lambda_0$ [see Eq. (2.2)] and minima for $s = \left(n + \frac{1}{2}\right)\lambda_0$, which is the condition for destructive interference.

Now, the mathematical form of the interferogram $I(s)$ is to be derived. The partial beams in the two arms of the Michelson interferometer (see Fig. 2) are described as plane waves. The amplitude or the electric field of the wave reflected at the fixed mirror is

$$E_1 = E_0\, e^{i(\omega t - \vec{q} \cdot \vec{r}_1)}, \qquad (2.3\text{a})$$

and the electrical field or the amplitude of the wave reflected at the movable mirror is

$$E_2 = E_0 \, e^{i(\omega t - \vec{q}, \vec{r}_2)} . \tag{2.3b}$$

The two waves are assumed to have equal intensity

$$I_0 = \frac{1}{2}\sqrt{\frac{\varepsilon_0}{\mu_0}}\, E_0^2 . \tag{2.4}$$

Here, intensity means the time average of the power flux per unit area, as follows from the theory of electrodynamics [32] $\left(I = \bar{S} = |\overline{\vec{E} \times \vec{H}}| \text{ and } \sqrt{\frac{\varepsilon_0}{\mu_0}} = 2.65 \cdot 10^{-3} \frac{\mathrm{A}}{\mathrm{V}}\right.$ if the units of $E$ are $\frac{\mathrm{V}}{\mathrm{m}}$ and those of $H$ are $\left.\frac{\mathrm{A}}{\mathrm{m}}\right)$. Superposition of the two waves yields the total intensity

$$\begin{aligned} I &= \frac{1}{2}\sqrt{\frac{\varepsilon_0}{\mu_0}}\, |E_1 + E_2|^2 \\ &= \frac{1}{2}\sqrt{\frac{\varepsilon_0}{\mu_0}}\, [E_0^2 + E_0^2 e^{-i\vec{q}, (\vec{r}_1 - \vec{r}_2)} + E_0^2 e^{+i\vec{q}, (\vec{r}_1 - \vec{r}_2)} + E_0^2] . \end{aligned} \tag{2.5}$$

Now we insert $I_0 = \frac{1}{2}\sqrt{\frac{\varepsilon_0}{\mu_0}}\, E_0^2$ and $q = \frac{2\pi}{\lambda_0} = 2\pi\tilde{\nu}_0$ ($\lambda_0$ and $\tilde{\nu}_0$ are again the wavelength and the wave number of the laser line, respectively) and take the path difference $s = \frac{1}{q}\, \vec{q}, (\vec{r}_1 - \vec{r}_2)$. The result is

$$I(s) = 2I_0[1 + \cos(2\pi\tilde{\nu}_0 s)] \tag{2.6}$$

for the interferogram of a single laser line. It should be noted that $I(s)$ is not identical with the signal at the detector but only proportional to it. In the following, however, this factor is suppressed in the interest of simplicity. Obviously, the function $I(s)$ given by Eq. (2.6) has maxima for $s = n\lambda_0$ ($I_{max} = 4I_0$) and minima for $s = \left(n + \frac{1}{2}\right)\lambda_0$ ($I_{min} = 0$), as expected from the interference conditions. The wavelength $\lambda_0$ of the laser radiation can easily be determined from the path difference $\Delta s$ between two maxima or two minima of $I(s)$.

Next, we consider the case of two narrow lines at wave numbers $\tilde{\nu}_1$ and $\tilde{\nu}_2$. The interferogram obtained in this case (see Fig. 2) is the superposition (sum) of the two interferograms of the lines. It exhibits the typical beat pattern usually encountered in connection with acoustical or electrotechnical problems. Mathematically, the interferogram is

$$I(s) = 2I_1[1 + \cos(2\pi\tilde{\nu}_1 s)] + 2I_2[1 + \cos(2\pi\tilde{\nu}_2 s)] \tag{2.7a}$$

or, under the assumption $I_1 = I_2 = I$ and after some rearrangement (with the use of the addition theorem of the angular functions),

$$I(s) = 4I\left[1 + \cos\left(2\pi\frac{\tilde{\nu}_1 - \tilde{\nu}_2}{2}s\right) \cdot \cos\left(2\pi\frac{\tilde{\nu}_1 + \tilde{\nu}_2}{2}s\right)\right] . \tag{2.7b}$$

The second cos factor describes the oscillation at the average wave number $\tilde{\nu} = \frac{1}{2}(\tilde{\nu}_1 + \tilde{\nu}_2)$ and the first cos factor is responsible for the beat pattern with $\Delta\tilde{\nu} = \frac{1}{2}(\tilde{\nu}_1 - \tilde{\nu}_2)$. Thus, the sum of the two wave numbers, or of the two frequencies, is obtained from the path difference $\Delta s$ between two maxima of the oscillation. The difference is obtained from the path difference between two maxima or two minima of the beat pattern. These two values are sufficient to determine $\tilde{\nu}_1$ and $\tilde{\nu}_2$.

However, this method of direct inspection of the interferogram does not work for more than two lines. For three lines of different intensities the interferogram is again the superposition of the interferograms of the single lines (see Fig. 2):

$$I(s) = 2\sum_{n=1}^{3} I_n[1 + \cos(2\pi\tilde{\nu}_n s)] , \tag{2.8}$$

but there is no simple way of extracting the frequencies and intensities of the three lines, *i.e.* the spectrum $I(\tilde{\nu})$, from this interferogram. Here we are at a crucial point for Fourier transform spectroscopy. How can we obtain the spectrum $I(\tilde{\nu})$ from the interferogram $I(s)$? The quantity we would like to obtain in a spectroscopic investigation is the spectrum $I(\tilde{\nu})$ and not the interferogram $I(s)$, and even in the simple case of three narrow lines there is no simple solution to this problem.

## 2.2 Fourier Transform

The problem is how to convert the interferogram $I(s)$ obtained with a Michelson interferometer into the spectrum $I(\tilde{\nu})$. Problems of this kind are met with in many areas of physics and technology [15], for example, the problem of determining the spectrum of harmonics for a musical instrument (flute or violin). At audiofrequencies the problem is easily solved with an appropriate set of electronic circuits that performs a so-called Fourier analysis. In Fourier transform spectroscopy the solution is obtained by mathematical treatment of the interferogram $I(s)$. In order to illustrate the principle of this treatment in a simple way, let us go back to the case of a single narrow laser line, *i.e.* monochromatic radiation.

From the interferogram $I(s)$ [see Eq. (2.6)] we take the oscillatory part after subtracting the average $2I_0$:

$$\bar{I}(s) = I(s) - 2I_0 = 2I_0 \cos(2\pi\tilde{\nu}_0 s) \tag{2.9}$$

and multiply by $\cos(2\pi\tilde{\nu}s)$, where $\tilde{\nu}$ may have any value between 0 and $\infty$ ($0 \leq \tilde{\nu} \leq \infty$). The curves so obtained are shown in Fig. 3 versus path difference $s$ for some selected values of $\tilde{\nu}$ ($\tilde{\nu} = 0{,}5\tilde{\nu}_0$; $\tilde{\nu} = 0.8\tilde{\nu}_0$; $\tilde{\nu} = \tilde{\nu}_0$; $\tilde{\nu} = 1.25\tilde{\nu}_0$; and $\tilde{\nu} = 1.5\tilde{\nu}_0$). Generally, the area under these curves vanishes, as the positive and negative contributions are equal. Only in the case $\tilde{\nu} = \tilde{\nu}_0$ are there exclusively positive contributions so that the area under the curve does not vanish (see Fig. 3). Mathe-

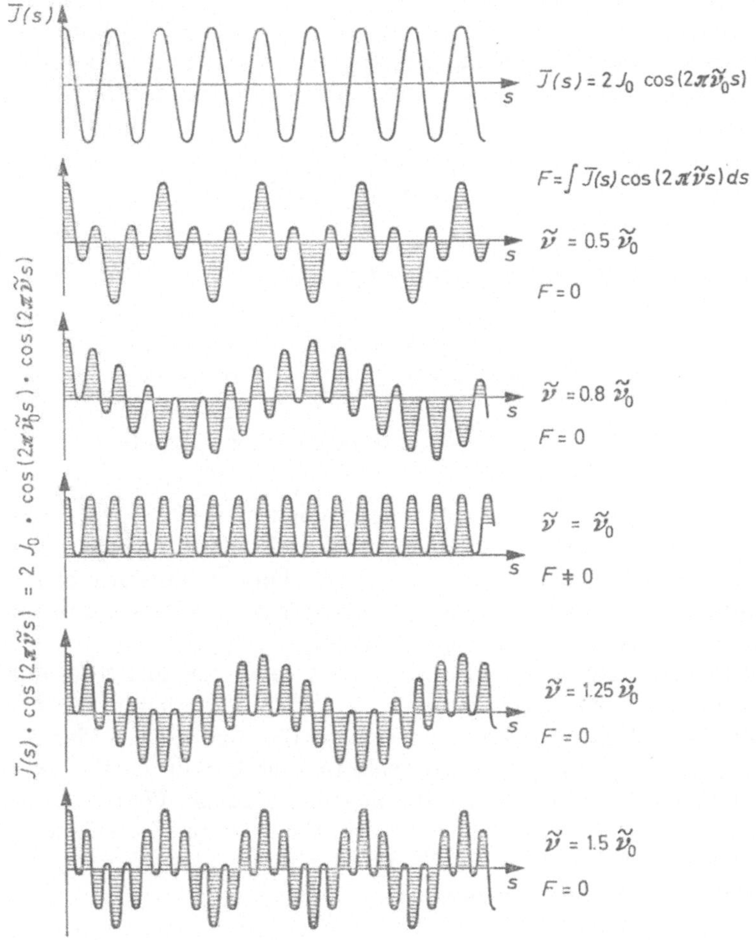


Fig. 3. Interferogram $\bar{I}(s)$ of a single narrow line versus path difference $s$, and the functions $\bar{I}(s)\cdot\cos(2\pi\tilde{\nu}s)$ for $\tilde{\nu} = .5\tilde{\nu}_0$., $\tilde{\nu} = .8\tilde{\nu}_0$, $\tilde{\nu} = \tilde{\nu}_0$, $\tilde{\nu} = 1.25\tilde{\nu}_0$, and $\tilde{\nu} = 1.5\tilde{\nu}_0$. The area $F$ under the curves which correspond to these functions is shaded

matically, this area is denoted by an integral, which is a function of the wave number $\tilde{\nu}$ and is written $I(\tilde{\nu})$:

$$I(\tilde{\nu}) = \int_{-\infty}^{+\infty} \bar{I}(s)\cos(2\pi\tilde{\nu}s)\,ds = 2I_0\int_{-\infty}^{+\infty}\cos(2\pi\tilde{\nu}_0 s)\cos(2\pi\tilde{\nu}s)\,ds \begin{cases} = 0 \text{ if } \tilde{\nu} \neq \tilde{\nu}_0 \\ \neq 0 \text{ if } \tilde{\nu} = \tilde{\nu}_0 \end{cases}. \quad (2.10)$$

This integral is a Fourier transform. $I(\tilde{\nu})$ is the Fourier transform of $\bar{I}(s)$ and can be shown to represent the spectral distribution of the intensity [32,34]. The mathematical procedure according to Eq. (2.10) can be performed for all possible values of $\tilde{\nu}$, and we may conclude that the wave number $\tilde{\nu}$ for which the integral does not vanish is identical with the wave number $\tilde{\nu}_0$ of the laser radiation. In other words, this mathematical procedure is a tool to obtain the value $\tilde{\nu}_0$ or $\lambda_0$ as obtained already from the interference condition Eq. (2.2) for a single laser line.

In the case of three narrow lines, however, $I(\tilde{\nu})$ can only be obtained by the mathematical procedure. The oscillatory part of the interferogram is now [see Eq. (2.8)]

$$\tilde{I}(s) = 2 \sum_{n-1}^{3} I_n \cos(2\pi\tilde{\nu}_n s) \tag{2.11}$$

According to Eq. (2.10), $I(\tilde{\nu})$ is calculated as follows:

$$I(\tilde{\nu}) = \int_{-\infty}^{+\infty} \tilde{I}(s) \cos(2\pi\tilde{\nu}s)\, ds \quad \begin{cases} \neq 0, \sim I_1 & \text{if } \tilde{\nu} = \tilde{\nu}_1 \\ \neq 0, \sim I_2 & \text{if } \tilde{\nu} = \tilde{\nu}_2 \\ \neq 0, \sim I_3 & \text{if } \tilde{\nu} = \tilde{\nu}_3 \\ = 0 \text{ for all other values of } \tilde{\nu}. \end{cases} \tag{2.12}$$

The calculation shows that $I(\tilde{\nu})$ is only nonzero if $\tilde{\nu}$ is equal to the wave number of one of the three lines ($\tilde{\nu}_1$, $\tilde{\nu}_2$ or $\tilde{\nu}_3$) and that it is proportional to the intensity of this line. Thus we see that $I(\tilde{\nu})$, the Fourier transform of $I(s)$, is the spectrum of the light emitted by the source and is more or less identical with the spectrum obtained with a grating spectrometer.

If $I(\tilde{\nu})$ is other than a few discrete narrow lines, the tool to evaluate $I(\tilde{\nu})$ from $I(s)$ is the Fourier transform, where $I(s)$ is the interferogram measured with a two-beam (Michelson) interferometer. This is the fundamental idea of Fourier transform spectroscopy. We have left aside the question of whether the Fourier integral Eq. (2.12) exists and whether it is meaningful or not. For the mathematical requirements on $I(s)$, the reader is referred to the literature [34]. It is sufficient to say here that, for all physically and experimentally reasonable interferograms $I(s)$, these requirements are usually met.

## 2.3 Resolution and Instrument Line-Shape Function

Having commented on the fundamentals, we now have to emphasize certain properties in more detail. We next discuss resolution, *i.e.* the minimum difference in the wave numbers of two narrow lines that can still be seen as separate lines by the spectrometer. This leads on to the instrument line-shape function of a spectrometer.

We again start with the diffraction grating. Let us assume that the source emits a single narrow line of negligible line-width, and that the intensity in the focal plane is studied with a sufficiently sensitive detector. We obtain a curve of intensity versus $\sin\varphi_d$, as shown in Fig. 4. There are main maxima at $\sin\varphi_d - \sin\varphi_i = \frac{n\lambda_0}{g}$ in accordance with Eq. (2.1). Between two main maxima, $(N-1)$ zeros are found at

$$\sin\varphi_d - \sin\varphi_i = \left(n + \frac{k}{N}\right)\frac{\lambda_0}{g} \quad \text{with} \quad \begin{matrix} n = 0, \pm 1, \pm 2, \ldots \\ k = 1,2,3, \ldots \end{matrix} \tag{2.13}$$

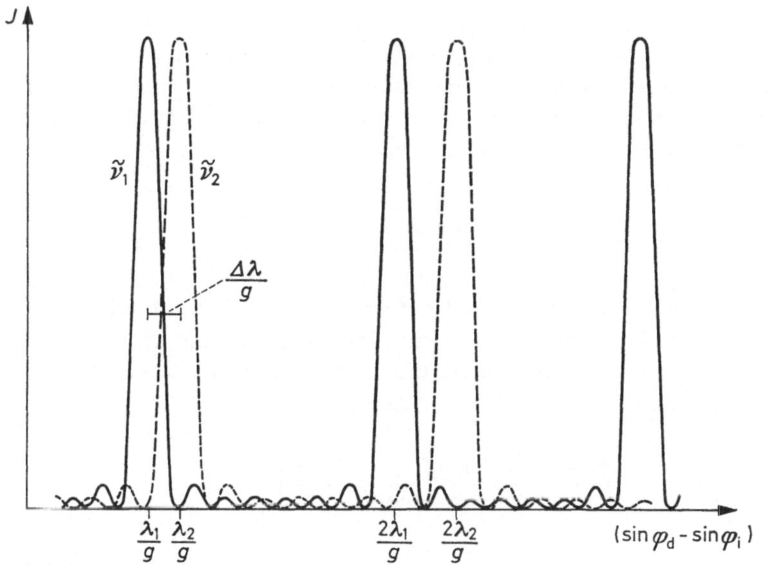


Fig. 4. Instrument line-shape function for a diffraction grating ($N = 8$) and the smallest difference $\Delta\lambda$ between two narrow lines which is clearly resolved by the grating

and also ($N$—2) secondary maxima [32,33]. Here $N$ is the number of lines in the grating. The mathematical formula for this intensity curve is usually derived by applying Huygens' principle [32]. The amplitude $E$ of the diffracted wave is the sum of the contributions generated at each periodicity interval of the grating by the incident wave:

$$E = C \cdot \sum_{n=1}^{N} e^{2\pi i \tilde{\nu} n g\,(\sin\varphi_d - \sin\varphi_i)} \tag{2.14}$$

where $C$ is proportional to the amplitude of the incident wave and depends on the shape of the lines. The intensity of the diffracted wave is thus proportional to

$$I \sim \frac{\sin^2[\pi N \tilde{\nu} g(\sin\varphi_d - \sin\varphi_i)]}{\sin^2[\pi\tilde{\nu} g(\sin\varphi_d - \sin\varphi_i)]}\,. \tag{2.15}$$

This factor represents the familiar line-shape function of a diffraction grating, some properties of which have been discussed above [see Eqs. (2.1) and (2.13) and Fig. 4]. The dependence of the factor $C$ on $\sin\varphi_d$ has been suppressed and consequently so has that of the intensity at the main peaks on $\varphi_d$. This depends strongly on the particular type of grating and may cause only a few diffraction orders to have a considerable intensity (*e.g.* echelette grating) [33].

The source is assumed to emit two narrow lines of nearly equal wave numbers $\tilde{\nu}_1$ and $\tilde{\nu}_2$, *i.e.* $(\tilde{\nu}_1 - \tilde{\nu}_2) \ll \frac{1}{2}(\tilde{\nu}_1 + \tilde{\nu}_2)$. Each line yields an intensity curve, as discussed in the last section (see Fig. 4). The smaller the difference $\Delta\tilde{\nu} = \tilde{\nu}_1 - \tilde{\nu}_2$, the more the main peaks of the intensity curves will penetrate each other until at $\Delta\tilde{\nu}=0$ they match. Thus, two neighboring spectral lines are usually said to be resolved, *i.e.* to give rise to clearly separated main peaks, if the main peak of one line coincides with the first zero of the other [32]. According to Eq. (2.13), the first zero, *i.e.* the zero next to a main maximum, occurs at

$$\sin\varphi_d - \sin\varphi_i = \left(n + \frac{1}{N}\right)\frac{\lambda_0}{g}\,. \tag{2.16}$$

From this relation and Eq. (2.1), it is apparent that the width of the main peak decreases as the number $N$ of lines increases. On the other hand, the ratio of the intensity of the main maximum to the intensity of the first secondary maximum does not depend on $N$ (for $N \gg 1$)

$$\frac{I\,(\text{main maximum})}{I\,(\text{1st secondary maximum})} = \left(\frac{3\pi}{2}\right)^2 \approx 22\,. \tag{2.17}$$

[The first secondary maximum occurs at $\sin\varphi_d - \sin\varphi_i = \left(n + \frac{3}{2N}\right)\frac{\lambda_0}{g}$ !]

After these explanations, we can express the condition for the resolution of two spectral lines as follows:

$$\begin{aligned} \sin\varphi_d - \sin\varphi_i &= n\,\frac{\lambda_1}{g} = \left(n + \frac{1}{N}\right)\frac{\lambda_2}{g} \\ \text{or}\quad \Delta\lambda = \lambda_1 - \lambda_2 &= \frac{\lambda_2}{nN} \approx \frac{\lambda}{nN} \end{aligned} \tag{2.18}$$

where $\lambda = \frac{1}{2}\,(\lambda_1 + \lambda_2)$ is the average of $\lambda_1$ and $\lambda_2$.

Conversion to wave numbers ($\tilde{\nu}=1/\lambda$, $\Delta\tilde{\nu}=\Delta\lambda/\lambda^2$) yields

$$\Delta\tilde{\nu} = \frac{\tilde{\nu}}{nN} = \frac{1}{nN\lambda}. \tag{2.19}$$

This is the well-known formula that states that the resolution of a diffraction grating increases (and $\Delta\tilde{\nu}$ decreases) with increasing order $n$ of diffraction and number $N$ of lines in the grating [33]. For the case of three lines, or for any other spectrum, the intensity is measured with a grating spectrometer as function of $\varphi_d$ (for several orders of diffraction) (see Fig. 5). The data obtained in this way are then easily converted to $I(\lambda)$ or $I(\tilde{\nu})$ and the problem of determining the spectral distribution $I(\tilde{\nu})$ is solved. It should be noted that the linewidth obtained (see Fig. 5) is influenced by the limited resolution of the instrument and that the linewidths of the three laser lines are assumed to be actually much smaller. In other words, we have been discussing the properties of the instrument line-shape function of a diffraction grating.

After this discussion of the resolution of a grating spectrometer, we proceed to the question of the resolution obtained with a Michelson interferometer. For

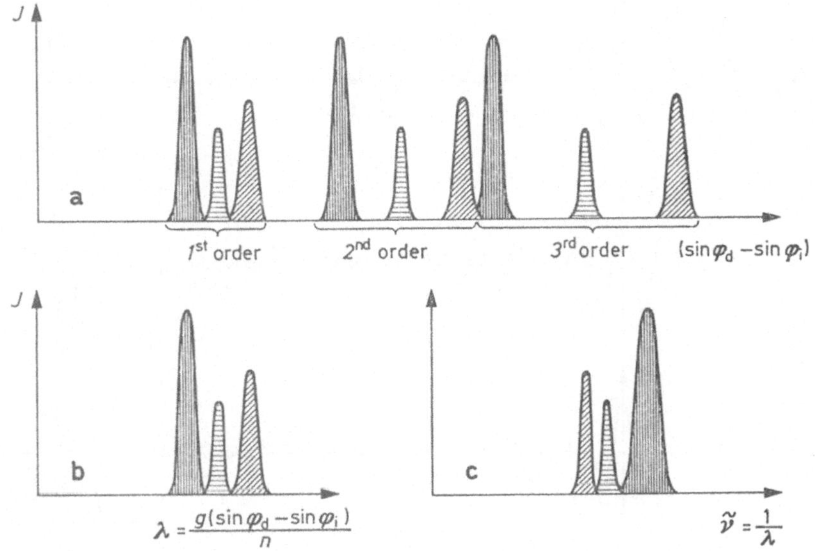


Fig. 5. Recording of the three narrow lines of different intensities by means of a diffraction grating. (The secondary maxima of the line-shape functions have been omitted here.) a) intensity versus ($\sin\varphi_d - \sin\varphi_i$); b) intensity versus wavelength $\lambda$; c) intensity versus wave number $\tilde{\nu}$

two narrow lines [see Fig. 1 and Eq. (2.7)], the difference $\Delta\tilde{\nu}$ in the wave numbers of the two lines governs the spacing of the beat pattern. We have to scan at least such maximum path difference $s_{max}$ to ensure that the first beat minimum is included in our scan. This is (see Fig. 1)

$$s_{max} = \frac{2}{\tilde{\nu}_1 - \tilde{\nu}_2} = \frac{2}{\Delta\tilde{\nu}} \quad \text{or} \quad \Delta\tilde{\nu} = \frac{1}{2 s_{max}} . \tag{2.20}$$

We can conclude from this that the resolution of a Michelson interferometer is proportional to the maximum path difference up to which the interferogram has been measured. When we now consider the case of three narrow lines, we must remember that we have to calculate $I(\tilde{\nu})$ from $I(s)$ by means of a Fourier transform [see Eqs. (2.10) and (2.12)]. However, the Fourier integral cannot be executed over $s$ from $-\infty$ to $+\infty$, since the interferogram $I(s)$ can be determined experimentally only over a finite range ($-s_{max} \leqq s \leqq +s_{max}$). Therefore, the integration too can be performed only over a finite range.

For the simplest case of a single laser line we obtain the observed intensity (Fig. 6)

$$\begin{aligned} I_{obs}(\tilde{\nu}) &= 2 I_0 \int_{-s_{max}}^{+s_{max}} \cos(2\pi\tilde{\nu}_0 s) \cos(2\pi\tilde{\nu} s)\, ds \\ &= 2 I_0 s_{max} \frac{\sin[2\pi(\tilde{\nu} - \tilde{\nu}_0) s_{max}]}{2\pi(\tilde{\nu} - \tilde{\nu}_0) s_{max}} \end{aligned} \tag{2.21}$$

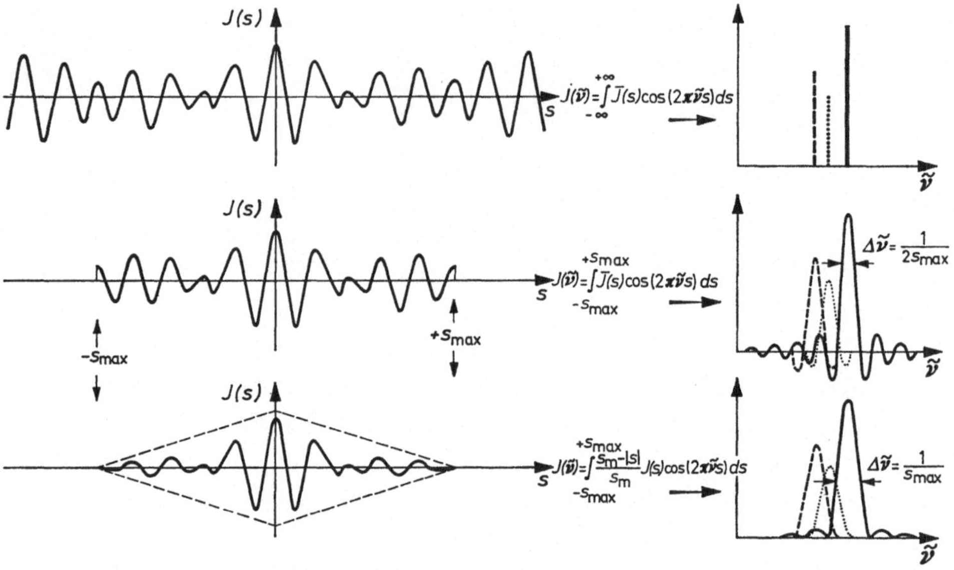


Fig. 6. Finite interferogram and resolution. (Example: Three narrow lines of different intensities); upper: infinite interferogram $I(\tilde{\nu})$ and corresponding spectrum; middle: finite interferogram and corresponding spectrum of the line-shape function $\left(\frac{\sin x}{x}\right)$; lower: finite interferogram with triangular apodization and corresponding spectrum of the line-shape function $\left(\frac{\sin x}{x}\right)^2$

where a similar term with $(\tilde{\nu}+\tilde{\nu}_0)$ has been neglected [35]. This function has a central main peak at $\tilde{\nu}=\tilde{\nu}_0$ with a half-width of approximately $\Delta\tilde{\nu}=\frac{1}{2s_{max}}$ [a]. These properties again express the fact that the resolution of a Michelson interferometer is proportional to the maximum path difference. Obviously, $I_{obs}(\tilde{\nu})$ is the analogue of the grating line-shape function [see Eq. (2.15) and Fig. 4]. Yet, $I_{obs}(\tilde{\nu})$ exhibits relatively large secondary extrema, positive and negative in sign, and the ratio between the intensity of the central maximum and that of the first secondary extremum (minimum) is much larger than in the case of the diffraction grating:

$$\frac{I\,(\text{central maximum})}{|I\,(\text{1st secondary extremum})|} = \frac{3\pi}{2} \approx 5\,. \tag{2.22}$$

The secondary extrema are disadvantageous for actual spectroscopic applications of the Michelson interferometer. They have their origin in the truncation, the

[a] The first zeros next to the main peak are located at $\tilde{\nu}-\tilde{\nu}_0=\pm\frac{1}{2s_{max}}$.

sharp and discontinuous cutoff of the interferogram. Therefore, the disadvantages will be diminished if the interferogram is forced to approach zero continuously for $s = \pm s_{\max}$. This can be achieved, for example, by multiplying $\bar{I}(s)$ by a triangular function (see Fig. 6). Then we obtain

$$I_{\text{obs}}(\tilde{\nu}) = 2I_0 \int_{-s_{\max}}^{+s_{\max}} \frac{s_{\max} - |s|}{s_{\max}} \cos(2\pi\tilde{\nu}_0 s) \cos(2\pi\tilde{\nu}s)\, ds = I_0 s_{\max} \frac{\sin^2[\pi(\tilde{\nu}-\tilde{\nu}_0)s_{\max}]}{[\pi(\tilde{\nu}-\tilde{\nu}_0)s_{\max}]^2} \tag{2.23}$$

where again a similar term for $(\tilde{\nu}+\tilde{\nu}_0)$ has been suppressed. The function given by Eq. (2.23) also has a central main peak at $\tilde{\nu}=\tilde{\nu}_0$ with a half-width of approximately $\Delta\tilde{\nu} = \frac{1}{s_{\max}}$ [b] [35]. In this case, the secondary maxima are relatively small in height. Actually, the intensity ratio is the same as for the diffraction grating [cf. Eq. (2.17)]

$$\frac{I(\text{central maximum})}{I(\text{1st secondary maximum})} = \left(\frac{3\pi}{2}\right)^2 \approx 22 \tag{2.24}$$

Multiplying the interferogram $\bar{I}(s)$ by a function, *e.g.* a triangular function, in order to make the product continuous at $s = \pm s_{\max}$ is called apodization [1–12,32]. The functions derived here are the line-shape functions of a Fourier spectrometer. If we compare the line-shape functions with apodization [see Eq. (2.21)] and without apodization [see Eq. (2.23)], we see that the apodization reduces the resolution by a factor of 2 but that many of the disadvantages due to the secondary extrema have been overcome. Here again, the clearly resolved difference in wave numbers $\Delta\tilde{\nu}$ is equal to that from the main peak to the first zero next to it in the line-shape function. Summarizing the results of this section, we refer to the effects of resolution and instrumental line-shape as shown in Fig. 6 for the case in which a spectrum of three narrow lines is analyzed by means of a Michelson interferometer. It should be noted that the choice of a triangular function for apodization is not the only possible one and that other functions could be selected, *e.g.* a cos function or the Genzel-Happ apodization (modified cos function) [26,36].

The resolving power of a spectrometer is usually defined as a dimensionless quantity

$$R = \frac{\tilde{\nu}}{\Delta\tilde{\nu}}. \tag{2.25}$$

For the diffraction grating, we obtain [see Eq. (2.19)]

$$R = nN = \tilde{\nu} s_{\max} \tag{2.26}$$

[b] The first zeros next to the main peak are located at $\tilde{\nu} - \tilde{\nu}_0 = \pm \frac{1}{s_{\max}}$.

where $s_{max} = nN\lambda$ is the maximum path difference between the elementary waves generated at the lines of the grating (see Fig. 1). For the Michelson interferometer, the resolution is

$$R = \tilde{\nu} s_{max} \tag{2.27}$$

when apodization is used. Eqs. (2.26) and (2.27) again express the similarity of the instrumental line-shape functions of diffraction grating and Michelson interferometer.

## 3. Spectroscopy

In the foregoing sections, the physical and optical fundamentals of spectroscopy have been considered, especially those of Fourier transform spectroscopy. In order to have simple interpretations, the source always was assumed to be a laser emitting one or more very narrow spectral lines. The sources used for practical spectroscopic applications, however, emit a continuous spectrum in the far-infrared spectral region (*e.g.* a high-pressure mercury lamp). In this section, therefore, the considerations of spectroscopy are extended to the case of continuous spectra, again starting with conventional (grating) spectroscopy and continuing with Fourier transform spectroscopy.

### 3.1 Conventional Spectroscopy

In this context, we are interested in an example of conventional spectroscopy only for comparison with Fourier transform spectroscopy; it is not intended to give a complete review of this field [33]. In order to be more specific, let us consider a grating spectrometer with a Czerny-Turner monochromator [37,38] as often used in far-infrared spectroscopy (see Fig. 7). The (continuous) spectrum emitted by the source is separated into its spectral elements by the diffraction grating. Each element can be characterized by a wave number $\tilde{\nu}$ (or wavelength $\lambda$) and a width $\Delta\tilde{\nu}$ (or $\Delta\lambda$), the radiation belonging to one of these elements being transmitted through the exit slit to the detector. This means that only a small portion of the intensity from the source reaches the detector. Turning the grating allows the intensity of the spectral elements $I(\tilde{\nu})$ to be recorded as a function of $\tilde{\nu}$ or $\lambda$ (see Fig. 8). Careful filtering is necessary to prevent overlap of neighboring orders of diffraction. However, filtering not only removes the unwanted radiation but also considerably diminished the wanted portion of the spectrum.

For sufficiently small widths of entrance and exit slits, the instrument line-shape function would be that of the diffraction grating (see Fig. 8). In the case of a continuous spectrum, the effect of the line-shape function and of the finite resolution is that each spectral element of infinitesimally small width produces such a line-shape function, and the recorded spectrum $I_{obs}(\tilde{\nu})$ is the superposition of all these. This means in practice that the ideal spectrum $I(\tilde{\nu})$ is scanned with this function or "spectral window" and that $I_{obs}(\tilde{\nu})$ contains contributions from the range $\tilde{\nu} \pm \Delta\tilde{\nu}$ (see Fig. 8). Therefore, it is often called scanning function or window

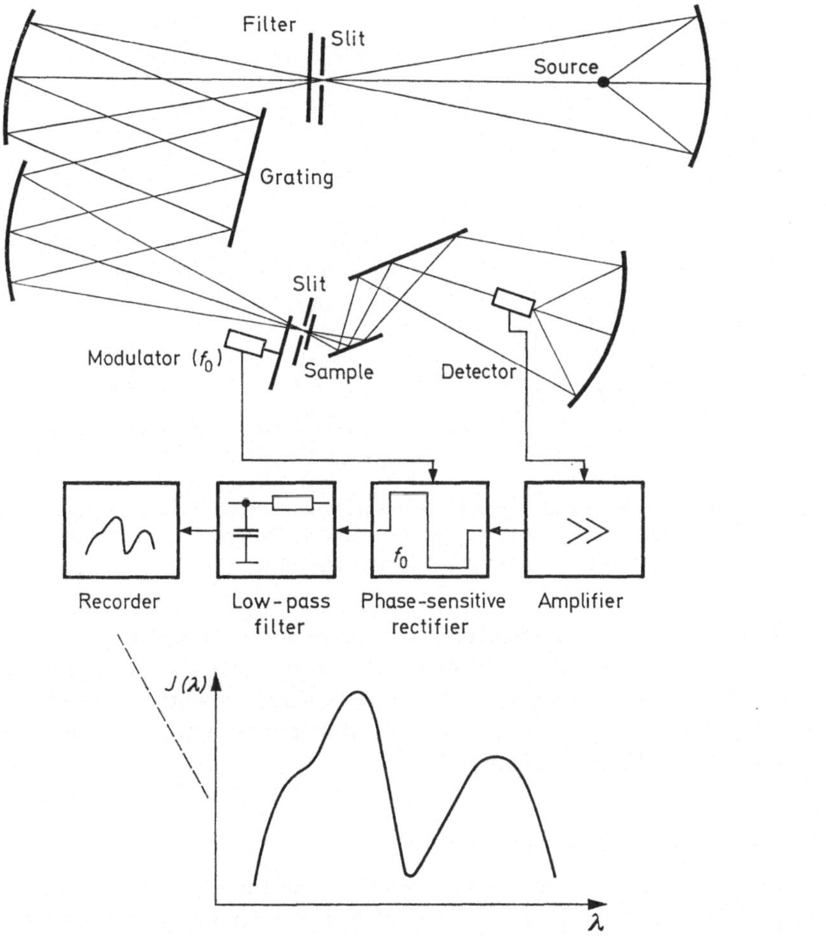


Fig. 7. Diagram of a grating spectrometer with the optical components (Czerny-Turner mount and the electronics with the usual lock-in technique

function [1–12]. Under the idealized conditions assumed here, the resolution would be $\Delta\tilde{\nu} = 0.05\ \text{cm}^{-1}$ at $\tilde{\nu} = 50\ \text{cm}^{-1}$ (or $R = 1000$) for a diffraction grating with $g = 0.3$ mm and $N = 1000$.

In practice, however, the slits have to be opened far enough to ensure that the energy reaching the detector is sufficient for a tolerable signal-to-noise ratio, *i.e.* for a meaningful measurement. The lack of more powerful sources and more sensitive detectors in the far-infrared has already been pointed out. Thus, the resolution of a grating spectrometer is not usually that resulting from the diffraction grating but by that from the slit width. Without entering into the details of the calculation here (see Section 5.1), we quote some typical figures for a grating spectrometer [39]:

$$\Delta\tilde{\nu} = 0{,}2\ \text{cm}^{-1} \quad \text{at} \quad \tilde{\nu} = 50\ \text{cm}^{-1} \quad \text{or} \quad R = 250\,. \tag{3.1}$$

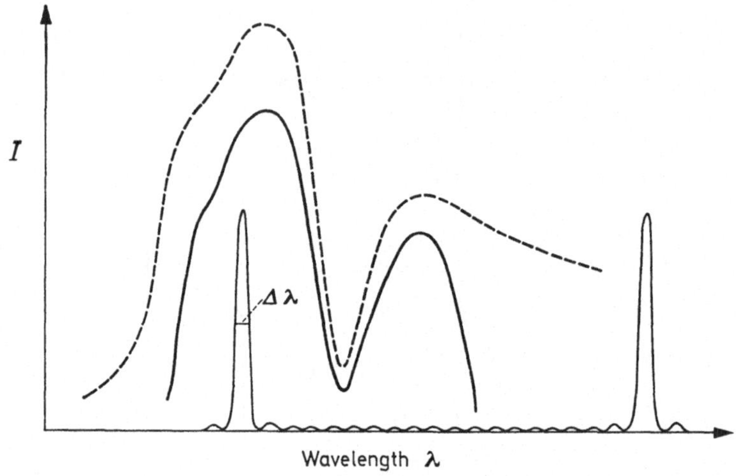


Fig. 8. Continuous spectrum $I$ versus wavelength $\lambda$ as recorded with a grating spectrometer (——) and the line-shape function (spectral window) for two orders of diffraction. For comparison, a spectrum with insufficient filtering is shown (- - - -)

In spectroscopic investigations, the quantities of interest are the transmission or reflection of the sample. In order to obtain these, the background spectrum (without sample in the spectrometer) and the sample spectrum have to be measured (Fig. 9), and their ratio calculated. In commercial double-beam instruments, this ratio is obtained automatically.

### 3.2 Fourier Transform Spectroscopy

Turning back to the Michelson interferometer and Fourier transform spectroscopy, let us first consider the interferogram of a continuous spectrum. Each spectral element of infinitesimal width $d\tilde{\nu}$ and intensity $I(\tilde{\nu})$ gives rise to the same interferogram pattern as a narrow line [see Eq. (2.6)], and the actual interferogram is the superposition of all these

$$I(s) = 2 \int_0^\infty I(\tilde{\nu})[1 + \cos(2\pi\tilde{\nu}s)]\, d\tilde{\nu} \,. \tag{3.2}$$

This interferogram (see Fig. 10 and Appdx 1) has some general properties:

a) $I(s)$ is symmetric about $s=0$ if the optical properties of the two arms of the Michelson are equal (which is assumed here).
b) Outside a certain path difference $s_c$ $(s > s_c)$ the interferogram is constant $I(s) = I(\infty)$ and independent of $s$. In this range, $s$ is greater than the coherence length of the light. There is no correlation between the two beams, and $I(s)$ is the sum of their intensities

$$I(\infty) = \int_0^\infty I(\tilde{\nu})\, d\tilde{\nu} + \int_0^\infty I(\tilde{\nu})\, d\tilde{\nu} \,. \tag{3.3}$$

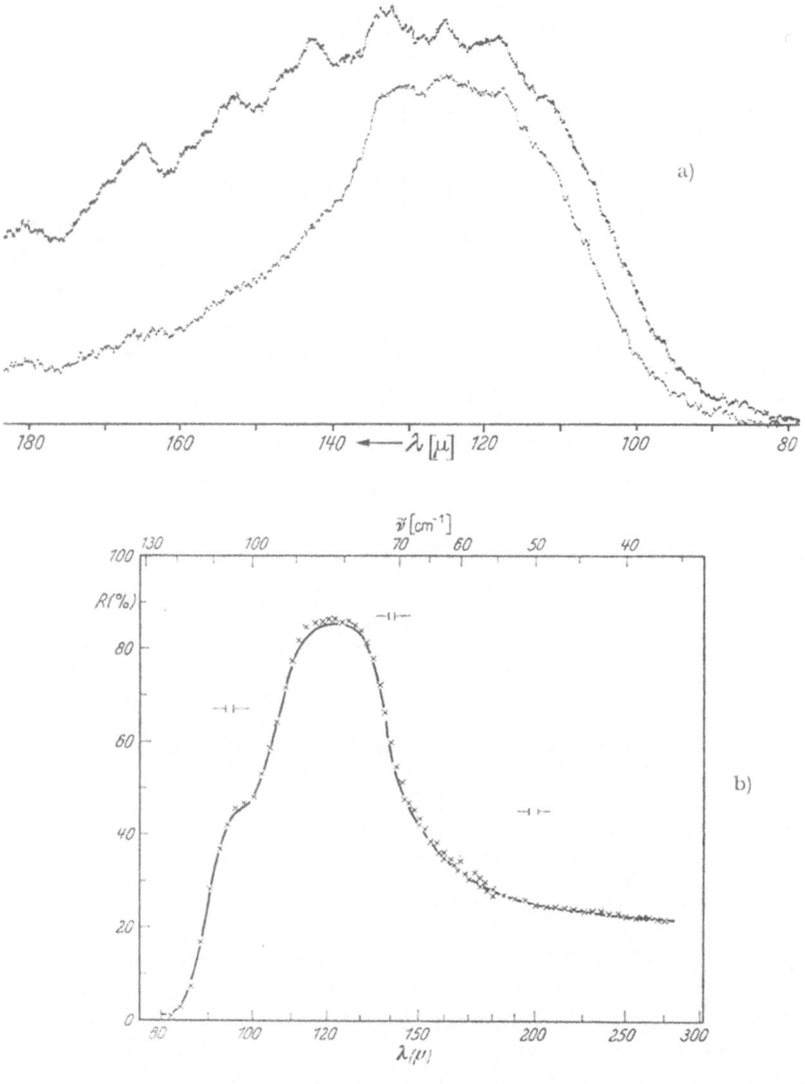


Fig. 9a and b. Spectra recorded by means of a grating spectrometer: a) background intensity (upper curve) and intensity with sample (lower curve) versus wavelength $\lambda$. b) ratio of the sample and background spectrum, in this example the reflectance of CsBr. The reflectance shown here was deduced not only from the experimental data in a) but also from other data. (Figures taken from Ref. [40])

c) The main central peak of the interferogram occurs at $s = 0$ where the light of all wavelengths is constructively interfering (white light position), and the following relation holds

$$I(0) = 2I(\infty) \tag{3.4}$$

[which follows from Eqs. (3.2) and (3.3)].

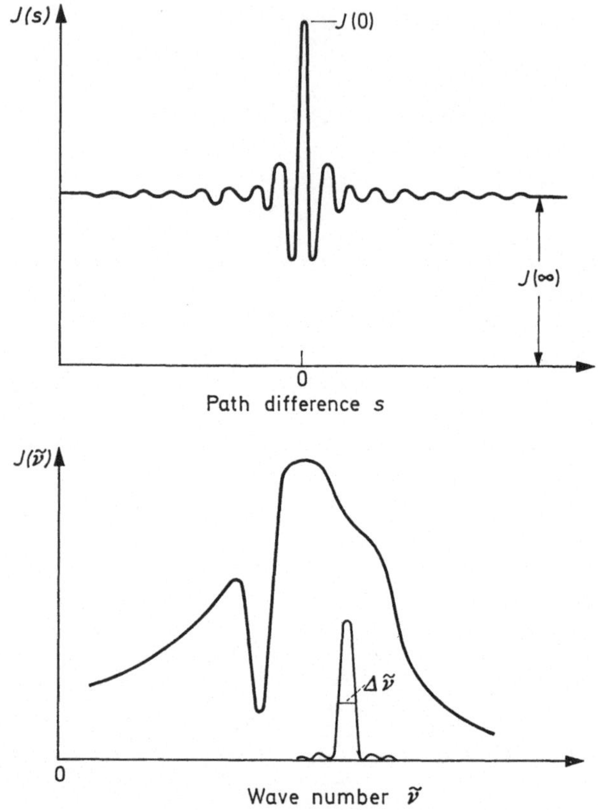


Fig. 10. Interferogram $I(s)$ of a continuous spectrum versus path difference $s$ (upper part), the corresponding spectrum $I(\tilde{\nu})$ versus wave number $\tilde{\nu}$, and the instrument line-shape function (for triangular apodization)

In practical applications of Fourier transform spectroscopy, such an interferogram [Eq. (3.2) and Fig. 10] is recorded by means of a two-beam interferometer. As in the case of a discrete spectrum with narrow lines, the spectrum $I(\tilde{\nu})$ is calculated from $I(s)$ by executing the Fourier transform. For this purpose, again only the oscillatory part of $I(s)$ is needed

$$\begin{aligned} \bar{I}(s) &= I(s) - \frac{1}{2} I(0) = I(s) - I(\infty) \\ &= 2 \int_0^\infty I(\tilde{\nu}) \cos(2\pi\tilde{\nu}s)\, d\tilde{\nu} \,. \end{aligned} \tag{3.5}$$

If we were able to determine $I(s)$ experimentally for all values of $s$ in the infinite range $-\infty \leqq s \leqq +\infty$, the result of the Fourier transform would be (see Appdx. 1)

$$I(\tilde{\nu}) = \int_{-\infty}^{+\infty} \bar{I}(s) \cos(2\pi\tilde{\nu}s)\, ds \,. \tag{3.6}$$

On comparing Eqs. (3.5) and (3.6), we realize that they are complementary to each other and that Eq. (3.5) represents a Fourier transform inverse to that in Eq. (3.6).

In practice, however, $I(s)$ is known only for a limited range: $-s_{\mathrm{max}} \leqq s \leqq s_{\mathrm{max}}$. For the reasons discussed already in Section 2.3, the interferogram $\bar{I}(s)$ is multiplied by a scanning or apodization function $S(s)$, *e.g.* $S(s) = \frac{s_{\mathrm{max}} - |s|}{s_{\mathrm{max}}}$ for $-s_{\mathrm{max}} \leqq s \leqq +s_{\mathrm{max}}$ and zero otherwise. We recall that the Fourier transform yields in the case of a narrow line at $\nu'$ the corresponding line-shape function $S(\tilde{\nu} - \tilde{\nu}') + S(\tilde{\nu} + \tilde{\nu}')$ where for our example [see Eq. (2.23)]

$$S(\tilde{\nu}) = s_{\mathrm{max}} \frac{\sin^2(\pi\tilde{\nu}s_{\mathrm{max}})}{(\pi\tilde{\nu}s_{\mathrm{max}})^2}.$$

In the case of a continuous spectrum, each spectral element at $\tilde{\nu}'$ with intensity $I(\tilde{\nu}')$ gives rise to a term $I(\tilde{\nu}')[S(\tilde{\nu} - \tilde{\nu}') + S(\tilde{\nu} + \tilde{\nu}')]$. Accordingly, the observed spectrum is ($I(\tilde{\nu}) = I(-\tilde{\nu})$!)

$$\begin{aligned} I_{\mathrm{obs}}(\tilde{\nu}) &= \int_0^{\infty} I(\tilde{\nu}')[S(\tilde{\nu} - \tilde{\nu}') + S(\tilde{\nu} + \tilde{\nu}')]\,d\tilde{\nu}' \\ &= \int_{-\infty}^{\infty} I(\tilde{\nu}')S(\tilde{\nu} - \tilde{\nu}')\,d\tilde{\nu}'. \end{aligned} \tag{3.7}$$

The physical meaning of Eq. (3.7) is that the true spectrum $I(\tilde{\nu}')$ is scanned with a line-shape function or spectral window $S(\tilde{\nu} - \tilde{\nu}')$ as in the case of the diffraction grating (see Fig. 10). As mentioned already, in contrast to the grating spectrometer, the spectral window can be varied according to the choice of the apodization function. The advantage of apodization is easily seen for a narrow laser line (cf. Fig. 6).

But what does apodization mean for a continuous spectrum? In a computer-simulated example (see Appdx) and where the ideal spectrum is known, the effect of the apodization is shown for two different values of $s_{\mathrm{max}}$ (Fig. 11). Without apodization, the secondary extrema of the spectral window [see Fig. 6 and Eq. (2.21)] produce undulations in the observed spectrum. With triangular apodization [see Fig. 6 and Eq. (2.23)], the undulations are absent. However, the resolution of the spectrum is considerably worse, as expected.

Although the spectrum $I_{\mathrm{obs}}(\tilde{\nu})$ is the required quantity from which the optical properties of the sample are to be deduced, it may be useful to know something about the "information" stored in the interferogram. This is demonstrated in Fig. 12 which shows separately the contributions to the interferogram that originate from a broad background spectrum and from an absorption line of smaller width (see Appdx). Near $s = 0$, the total interferogram is dominated by the interference patterns from the broad background, which decay rather rapidly with increasing path difference $s$. The interferogram is then governed by the influence of the absorption line. In this example, the background spectrum as well as the absorption line are of "Lorentzian" type, and the interferograms (the Fourier transform) are damped cosine waves. As the curves in Fig. 12 show, the broader the spectrum, the greater the damping of the cosine wave in the interferogram. This statement

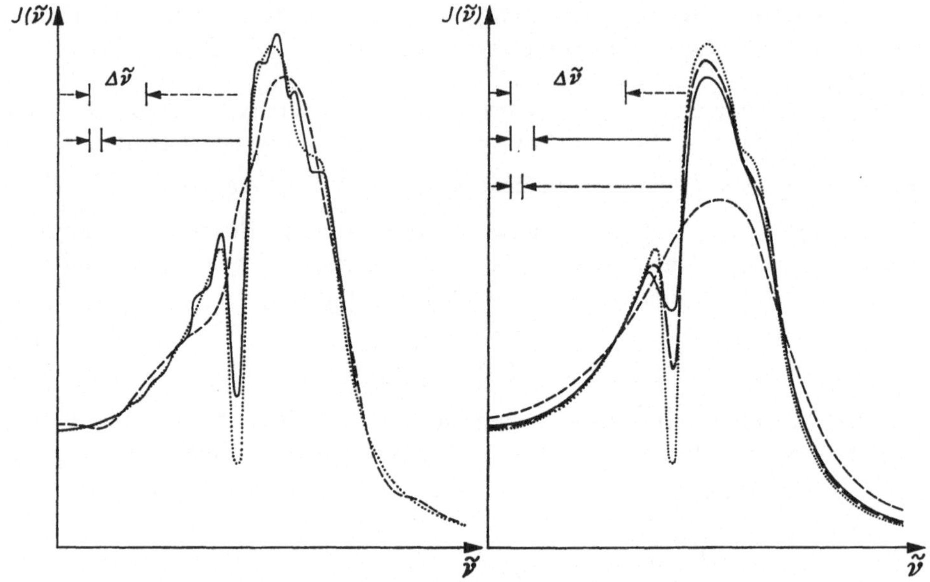


Fig. 11. Comparison of spectra obtained by means of Fourier transform spectroscopy with (right) and without (left) apodization for low ( - - ), medium (——) and high (— —) resolution. The "true" spectrum for infinite resolution (. . . . .) is also shown in this computer-simulated demonstration

is not restricted to the example given here but applies quite generally to most of the Fourier transform pairs of functions. For "reading" and interpreting certain details of experimentally recorded interferograms, the reader is recommended to familiarize himself with a number of standard Fourier transform pairs quoted here or elsewhere [15], then it is not too difficult to deduce some details of the spectrum from the interferogram.

We now arrive at the conclusion that in spite of all the differences there are many similarities between Fourier transform and conventional spectroscopy. For example, the instrument line-shape functions are very similar for triangular apodization and for a diffraction grating, but one difference between them is that the line-shape function in Fourier spectroscopy has only one central main peak while there are many orders of diffraction in case of the grating. However, this difference exists only if the Fourier transform is executed in an analogous way. If the Fourier transform is performed by means of a digital computer, the effective line-shape function also has an infinite number of main peaks. The reason for this is that the use of a digital computer requires a digitizing of the interferogram. This is usually done by recording the interferogram at equal increments $\Delta s$ of optical path difference. Instead of the continuous function $I(s)$ a finite series of equally spaced interferogram points $I(n\Delta s)$ is obtained (Fig. 13).

Now, the procedure of the Fourier transform is no longer an integral but a sum

$$I_{\text{obs}}(\tilde{\nu}) = \sum_{n=-\infty}^{+\infty} \bar{I}(n\Delta s) \cos(2\pi\tilde{\nu}n\Delta s)\,\Delta s\,. \tag{3.8}$$

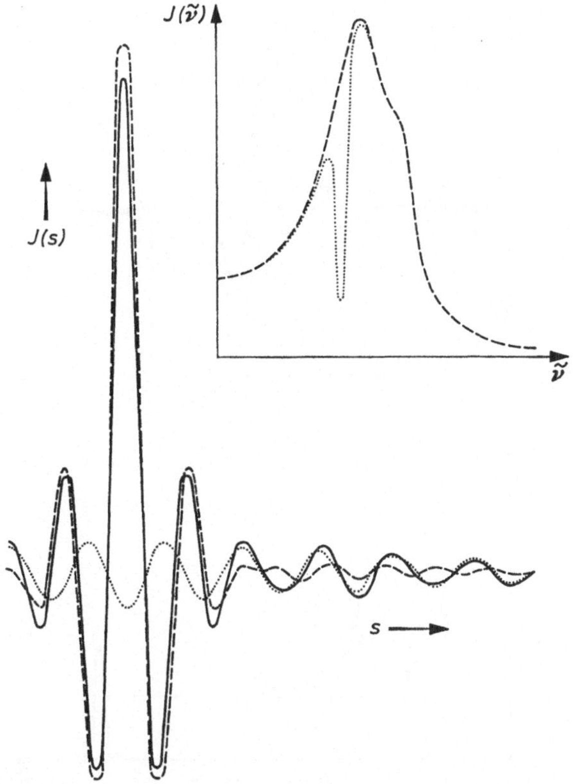


Fig. 12. Illustration of the information in an interferogram: broad background spectrum and its interferogram (- - -); narrow absorption line and its interferogram (. . . .). The sum of the two contributions yields the total interferogram (——)

It can be proved mathematically that $I_{obs}(\tilde{\nu})$ is identical with the true spectrum $I(\tilde{\nu})$ if this is nonzero only in $0 \leqq \tilde{\nu} \leqq \tilde{\nu}_{max} < \frac{1}{2\Delta s}$ and if $\tilde{\nu}$ is restricted to this range. Furthermore, it is assumed that the interferogram is known for $-\infty \leqq n \leqq +\infty$ [4, 15]. When, instead of $\tilde{\nu}$ (in the range $0 \leqq \tilde{\nu} \leqq \tilde{\nu}_{max}$), the wave number $\tilde{\nu} + m\frac{1}{\Delta s}$ outside the above range is inserted in Eq. (3.8), it is easily seen that

$$I_{obs}\left(\tilde{\nu} + m\frac{1}{\Delta s}\right) = I_{obs}(\tilde{\nu}) \qquad m = 0, \pm 1, \pm 2, \ldots . \tag{3.9}$$

due to the periodicity of the cosine function. Physically, this means that the true spectrum $I(\tilde{\nu})$ is reduplicated a great number of times equally spaced by $\Delta\tilde{\nu} = m\frac{1}{\Delta s}$ on the $\tilde{\nu}$ scale (see Fig. 13), like the diffraction orders of a grating. This can be expressed as follows:

$$I_{obs}(\tilde{\nu}) = \sum_{m=-\infty}^{+\infty} I\left(\tilde{\nu} + m\frac{1}{\Delta s}\right) \tag{3.10}$$

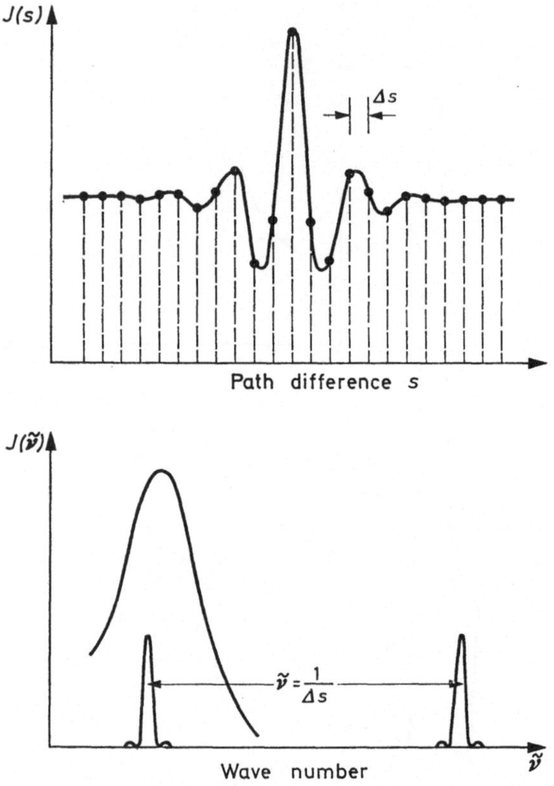


Fig. 13. Sampling of an interferogram at equal increments $\Delta s$ of the path difference (upper part) and the spectrum with instrument line-shape function (lower part)

or, if the effects of finite interferogram and apodization are included [see Eq. (3.7)]:

$$I_{\text{obs}}(\tilde{\nu}) = \sum_{m=-\infty}^{+\infty} \int_0^{\infty} I(\tilde{\nu}') \left[ S\left(\tilde{\nu} - \tilde{\nu}' + m\frac{1}{\Delta s}\right) + S\left(\tilde{\nu} + \tilde{\nu}' + m\frac{1}{\Delta s}\right)\right] d\tilde{\nu}' . \quad (3.11)$$

Now we can redefine the line-shape function

$$\hat{S}(\tilde{\nu}) = \sum_{m=-\infty}^{+\infty} S\left(\tilde{\nu} + m\frac{1}{\Delta s}\right)$$

and finally write

$$I_{\text{obs}}(\tilde{\nu}) = \int_0^{\infty} I(\tilde{\nu}') [\hat{S}(\tilde{\nu} - \tilde{\nu}') + \hat{S}(\tilde{\nu} + \tilde{\nu}')] d\tilde{\nu}' . \quad (3.12)$$

The new line-shape function $\hat{S}(\tilde{\nu})$ incorporates the influence of apodization etc. on the spectrum as well as that of the digitizing. Instead of one central peak it exhibits a number of main peaks (see Fig. 13), and it is this fact that introduces the problem of filtering in Fourier transform spectroscopy, a problem already encountered in grating spectroscopy.

In order to make $I_{obs}(\tilde{\nu})$ unique, it follows from Eq. (3.10) or (3.12) and from $I(-\tilde{\nu}) = I(\tilde{\nu})$ that $I(\tilde{\nu})$ must be nonzero only in the range $0 < \tilde{\nu} < \frac{1}{2\Delta s}$ $\left[\text{or in a range } \frac{m}{2\Delta s} < \tilde{\nu} < \frac{m+1}{2\Delta s}\right]$. Thus, we have to use a filter such that $I(\tilde{\nu}) = 0$ for $\tilde{\nu} \geqq \tilde{\nu}_{max}$, and then we have to choose the sampling interval $\Delta s \leqq \frac{1}{2\tilde{\nu}_{max}}$. It should be noted that these filter requirements are not nearly as severe as in the case of grating spectroscopy. If we are using the first order of the grating, the usable region has to be restricted to one octave. From 50 $cm^{-1}$ to 400 $cm^{-1}$, *i.e.* over three octaves, we have to change gratings and filters several times. In Fourier transform spectroscopy, this range may be scanned in one run with one filter removing unwanted radiation from above 400 $cm^{-1}$. The sampling interval has to be $\Delta s = \frac{1}{800}$ cm = 12.5 $\mu$m. These figures also indicate the precision with which the position of the movable mirror has to be measured. In the far-infrared, this is usually done by means of a Moiré system; in the middle- and near-infrared, the position of the mirror may be controlled more accurately with a laser system [41)].

With a Moiré system, sampling intervals as small as $\Delta s = 5$ $\mu$m are achieved with sufficient precision. For this value, the maximum wave number is $\tilde{\nu}_{max} =$ 1000 $cm^{-1}$, which is not far from those of the emission lines of a $CO_2$ laser ($\tilde{\nu} \approx 943$, 970, 1040, and 1075 $cm^{-1}$). In order to suppress the exciting laser line in laser Raman experiments, edge filters have been developed with high transmission in the region below 900 $cm^{-1}$. After that, there is a rather sharp decrease in transmission, and between 930 and 970 $cm^{-1}$ they become opaque. Filters of this kind are very useful in Fourier spectroscopy, too, when $\tilde{\nu}_{max} = 1000$ $cm^{-1}$. Fig. 14 shows an example of a spectrum obtained with satisfactory filtering together with a similar spectrum where the filters are inadequate. The original spectrum $I(\tilde{\nu})$ is in the region 0 to 1000 $cm^{-1}$; for the range 1000 to 2000 $cm^{-1}$, the digital Fourier transform yields the spectrum $I(-\tilde{\nu})$ shifted by $\frac{1}{\Delta s} = 2000$ $cm^{-1}$. For clarity, $I(-\tilde{\nu}) = I(\tilde{\nu})$ has also been drawn in Fig. 14. If the original spectrum $I(\tilde{\nu})$ extends beyond 1000 $cm^{-1}$ (inadequate filters), there is an overlap of $I(\tilde{\nu})$ and $I\left(-\tilde{\nu} + \frac{1}{\Delta s}\right)$, and from the results of the Fourier transform at 950 $cm^{-1}$, for example, it is impossible to decide whether the intensity obtained originates from radiation at 950 $cm^{-1}$ or from radiation at 1050 $cm^{-1}$. To make this demonstration of aliasing in Fourier transform spectroscopy complete, the center section of the interferogram for the satisfactorily filtered spectrum is also given in Fig. 14. The sampling interval has been indicated. The maximum path difference $s = 0.2$ mm as shown there would yield a rather poor resolution of $\Delta\tilde{\nu} = 50$ $cm^{-1}$. In fact, the spectra were computed from longer interferograms with better resolution.

For most of the Fourier spectrometers, the maximum feasible path difference is about $s_{max} = 20$ $cm^{-1}$. The equivalent value of the resolved difference in wave numbers is $\Delta\tilde{\nu} = 0.05$ $cm^{-1}$. This means that the resolving power had the value

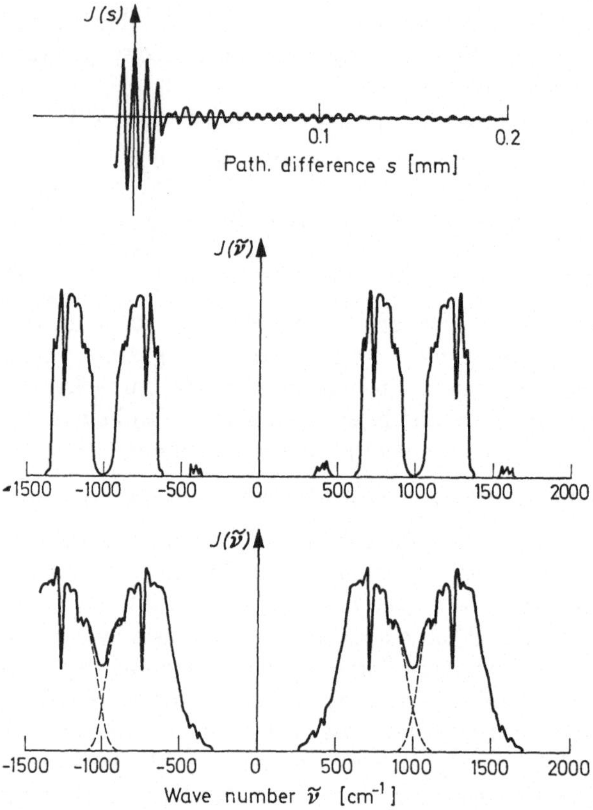


Fig. 14. Demonstration of sufficient (center part) and insufficient (lower part) filtering in Fourier transform spectroscopy. In addition to the spectra $I(\tilde{\nu})$, the interferogram $I(s)$ is shown for the spectrum with sufficient filtering (upper part). These data were obtained with a Polytec FIR 30 Fourier spectrometer

$R = 1000$ at $\tilde{\nu} = 50$ cm$^{-1}$. In practice, a good resolution can be achieved in Fourier transform spectroscopy without too much effort. It is easily increased by increasing the maximum path difference rather than by narrowing the slits and decreasing the luminous flux in the instrument; anyway, this flux is larger in a Michelson interferometer than in a grating instrument. A principal difference between the two methods is that in Fourier transform spectroscopy $\Delta\tilde{\nu}$ is constant for a measurement with given $s_{max}$ and a certain apodization. With a diffraction grating in n-th order, $R = \tilde{\nu}/\Delta\tilde{\nu}$ is constant for an investigation, provided sufficiently narrow slits can be used.

## 4. Practice of Fourier Transform Spectroscopy

So far, the theory of Fourier transform spectroscopy has been developed. This section is devoted more to the experimental and practical aspects. Naturally, comments on the practical realization start with the various kinds of two-beam interferometers which have been used for this purpose. Further, a thorough discussion seems necessary on the different methods of executing the Fourier transform in analogue or digital way. However, we shall not enter into the details of the computer program needed for the Fourier transform and shall consider mainly the principal ideas of the method. On the one hand, the computer program is provided by the manufacturer for commercial instruments; on the other hand, it is not too difficult to obtain programs from other people working in the field or from the literature [4]. However, in many cases these programs have to be adapted or modified for the computer to be used.

### 4.1 Two-Beam Interferometers

Of the two-beam interferometers employed for Fourier transform spectroscopy, the Michelson interferometer has already been mentioned several times (see. Fig. 2 and Section 2). Perhaps it is worth recalling here the fact that in most far-infrared spectrometers (conventional or Fourier) all optical elements are mirrors due to the lack of materials for lenses suitable for a wide range in the far infrared. In order to keep the optical setup compact, off-axis parabolic and elliptic mirrors are often employed in commercial instruments. However, the most important part of the Michelson interferometer is the beam splitter. Therefore, let us consider it in more detail. In the visible, near- and middle-infrared spectral regions the beam splitter usually is a dielectric film (or coating) on a transparent substrate. Materials like Ge and $Fe_2O_3$ are frequently used for the dielectric film, while $CaF_2$, KBr and quartz play a dominant roll as substrate materials, especially in commercially available instruments (see Section 6.2). A rather unusual beam divider is a reflection grating as employed in an all-reflecting-Michelson-interferometer [40] for the visible and near-infrared. In the far-infrared region on the other hand, no transparent substrate is available but, for the long waves, the dielectric film can be made much thicker. Thus self-supporting dielectric film beam splitters are more convenient in this region. Appropriate materials are sheets of organic polymers (Mylar, PET etc.)

The thickness of these beam dividers is comparable to the wavelength of the radiation, and they exhibit the interference phenomena typical for a thin film or a Fabry-Perot etalon. The reflectivity $R$ and transmission $T$ of the film are periodic functions of $d \cdot \tilde{\nu}$, where $d$ is its thickness and $\tilde{\nu}$ the wave number of the radiation (Fig. 15). Obviously, this periodicity is caused by the interferences within the film (channel spectra). The relative efficiency of the beam splitter is given by $4\,RT$; this quantity is unity for the ideal values $R = 0.5$ or 50%, and T = 0.5 or 50%; it is less than unity otherwise. Whether the optimum value $4\,RT = 1$ is achieved in practice, depends on whether $T$ and $R$ reach their ideal value or not (see Fig. 15), and this in turn depends on the refractive index $n$ of the film, on the angle of incidence, and the direction of polarization. Absorption can generally be neglected. The zeros of $4\,RT$ always occur when the optical path difference for an additional zigzag reflection in the film is an integral multiple of the wave length. From these arguments, it follows that for practical applications the most favorable thickness of the beam splitter has to be selected. Since the first zero of $4\,RT$ occurs at a certain value of $d \cdot \tilde{\nu}$, a thin beam splitter is suitable

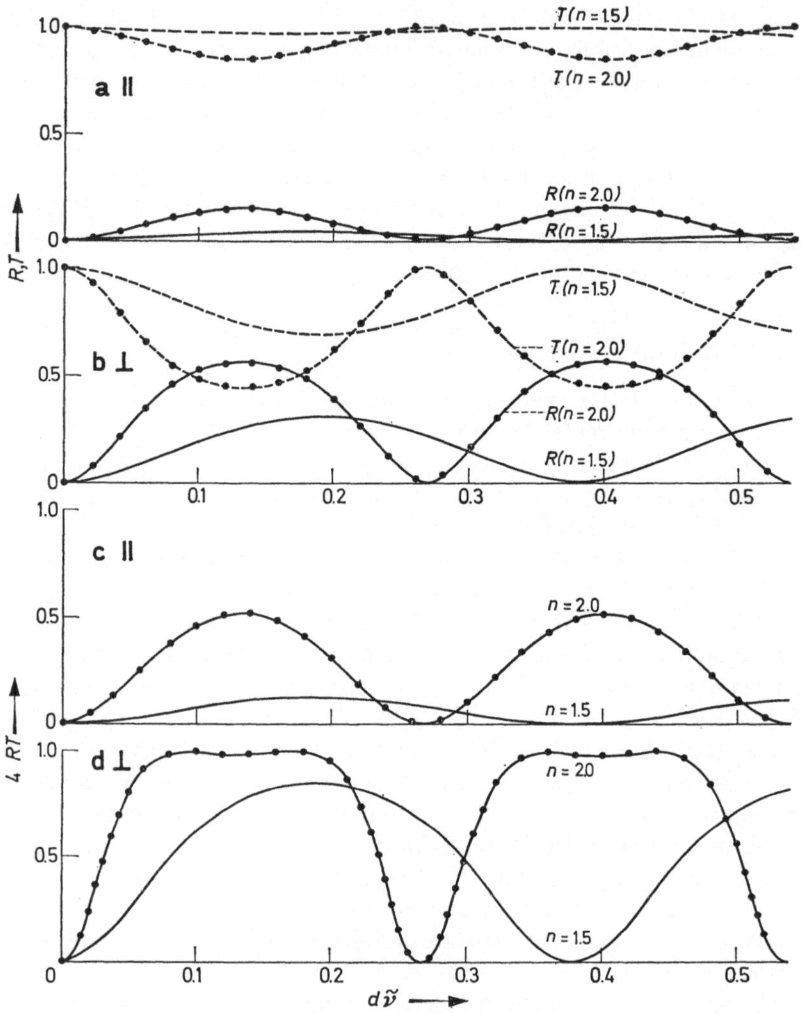


Fig. 15. Reflectance $R$ and transmittance $T$ (a, b) and beam splitter efficiency $4\ R\ T$ (c, d) versus $d\tilde{\nu}$ (thickness $d$ and wave number $\tilde{\nu}$) for thin films with refractive indices $n = 1.5$ and $n = 2.0$. The angle of incidence was assumed to be $\varphi = 45°$. The curves were evaluated for light polarized parallel to (a, c, ∥) and perpendicular to the plane of incidence (b, d, ⊥)

for a wide range up to relatively high wave numbers and a thick beam splitter for the very far-infrared (Fig. 16). Due to the limited range of efficiency of the beam dividers, all commercial Fourier spectrometers provide a set of them: various coatings on different substrates in the near- and middle-infrared, and mylar films with several thicknesses in the far-infrared, respectively (see Section 6.2). In some models for the far-infrared, the set of beamsplitters is mounted on a rotating wheel, and the change can be made automatically without breaking the vacuum of the instrument. It should also be pointed out in more detail that the angle of incidence of the radiation onto the beam divider is close to the Brewster

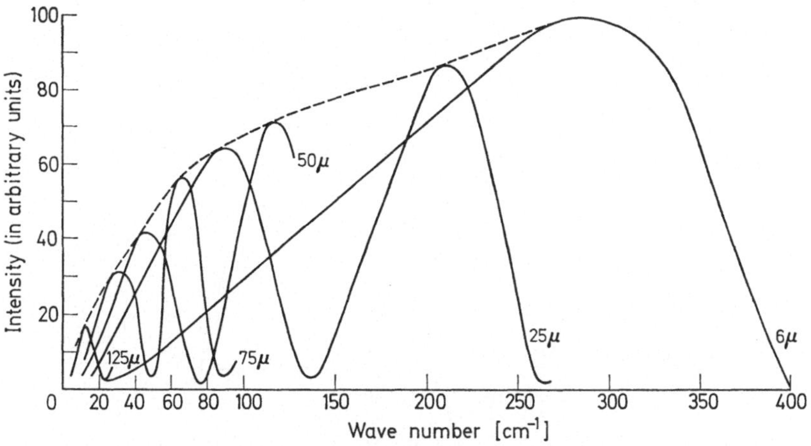


Fig. 16. Relative efficiency of thin-film mylar beam splitters with several thicknesses (experimentally determined) as a function of wave number. Taken from Ref. [42)]

angle in some far-infrared instruments. For an angle of 45° and $n = 1.5$ as an example, light polarized parallel to the plane of incidence, *i.e.* the plane of the optical diagram of the Michelson interferometer, is nearly 100% transmitted (cf. Fig. 15), and the beam splitter efficiency is rather poor in this case. In fact, the efficiency 4RT is about 7 times larger for the other polarization, *i.e.* perpendicular to the plane of incidence (cf. Fig. 15). This has to be taken into account for investigations of anisotropic samples with polarized light. In most of these cases, it is more advisable to turn not the polarizer by 90° but the sample in such a highly polarizing instrument when a measurement is to be performed for the other polarization. For the dielectric films on a substrate used as beam dividers in the middle- and near-infrared, similar relations hold as those plotted for self-supporting films in Fig. 15. However, the thickness of the dielectric coating is rather small, and the range of usability of these beamsplitters is wider (cf. Fig. 47 in Section 6.2). Also polarization problems are not so stringent in this case. Mostly, compensation plates are inserted in the radiation path in such a way that the asymmetry due to the substrate is compensated.

All thin-film beam splitters have the property that their efficiency tends to zero as $\tilde{\nu}^2$ if $\tilde{\nu}$ tends to zero. This is a great disadvantage for the extreme far-infrared. For the region at about 1 mm wavelength, therefore, another two-beam interferometer is employed with better success: the lamellar grating interferometer (Fig. 17). For the purpose of Fourier transform spectroscopy, the grating consists of two interpenetrating sets of facets or strip mirrors, one of which is movable and the other fixed. There is a path or phase difference between the rays reflected at the fixed set and those reflected at the movable set, and the interference of all these rays results in the pattern that would have been obtained with a Michelson interferometer. The path difference is dependent on the depth of the grating, *i.e.* the position of the movable strip mirror, and by varying its position, an interferogram $I(s)$ may be scanned as a function of path difference $s$.

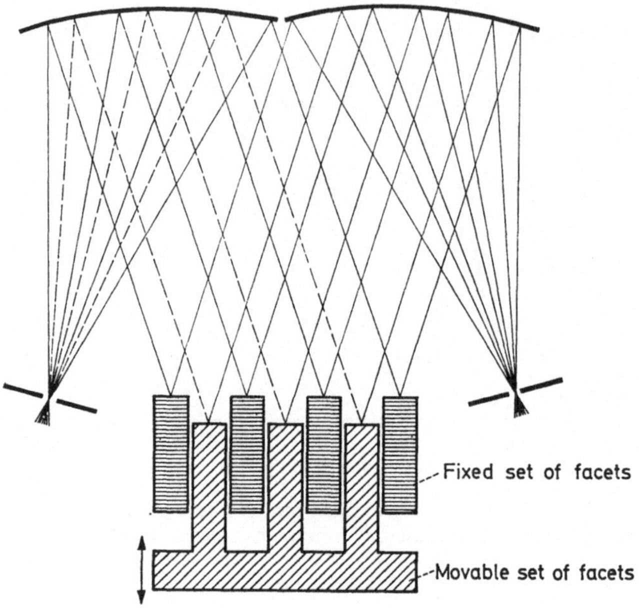


Fig. 17. Lamellar grating with two interpenetrating sets of facets

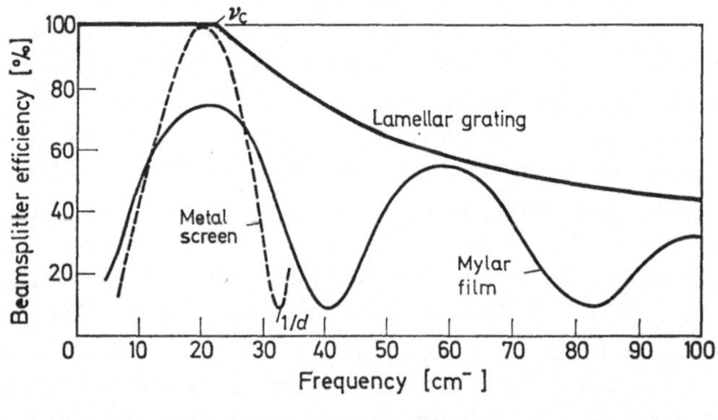


Fig. 18. Two-beam interferometer efficiencies of a Michelson interferometer with a thin-film mylar beam splitter (—), with a metal screen as beam splitter (- -), and with a lamellar grating (——). These data were taken from Ref. [44]

From the point of view of diffraction, the radiation in zeroth order is transmitted by the exit slit to the detector. The efficiency of this interferometer is rather high for long waves (Fig. 18). There is a high frequency cutoff due to cancellation of the interferences when the first order of diffraction also enters the exit slit. This cutoff can easily be calculated from the grating formula Eq. (2.1) [nearly normal incidence, $\varphi_d \approx \varphi_i \approx 0$, 1st order]:

$$\lambda = g(\sin \varphi_d - \sin \varphi_i) \approx g(\varphi_d - \varphi_i) \approx g\frac{w}{2f}$$

where $w$ is the exit slit width and $f$ the focal length of the system [$g$ is again the grating constant]. There is also a low frequency cutoff for the lamellar grating interferometer when the wavelength becomes comparable to the grating constant. For large path differences, *i.e.* large resolutions, problems may arise from shadowing.

In order to achieve better performance for the Michelson interferometer, metal meshes and wire grids have also been used as beam dividers. An especially interesting solution of this problem is the polarizing interferometer as developed by D. H. Martin [45], where good efficiency is obtained over a wide spectral range. In addition, the recorded signal is $\bar{I}(s)$, *i.e.* only the oscillatory part of the interferogram. This method avoids the difficulty that spurious features are produced in $I_{obs}(\tilde{\nu})$ because of a shift of the mean level $I(\infty)$ due to instrumental instabilities.

The mean level also drops out when, in addition to the transmitted interferogram, the interferogram reflected back to the source is recorded and the difference of the two interferograms is formed. For such a procedure, two detectors have been used in the double output interferometer by W. J. Burrough *et al.* [47]. In order to understand the underlying basic idea, it is necessary to recall that the total intensity of electromagnetic radiation sent by the source into the Michelson-interferometer is [see Eqs. (2.3–2.6) and Eq. (3.2)]

$$4\int_0^\infty I(\tilde{\nu})\, d\tilde{\nu}\,.$$

When the path difference s is large, the portion transmitted by an ideal Michelson interferometer is [see Eq. (3.3)]

$$I(\infty) = 2\int_0^\infty I(\tilde{\nu})\, d\tilde{\nu}$$

and the same amount of radiation energy is reflected back to the source. For smaller path differences s where interferences between the two partial beams are observed, the intensity reflected back to the source is

$$I_{\mathrm{Refl}}(s) = 2\int_0^\infty I(\tilde{\nu})\,[1 - \cos(2\pi\tilde{\nu}s)]\, d\tilde{\nu}\,. \tag{4.1}$$

Eq. (4.1) can be derived in the same way as Eq. (3.2) for the transmitted intensity was derived. Since we assumed an ideal instrument with no losses and no sample in the radiation path, we need not enter into mathematical details. From the requirement of energy conservation it follows immediately that the sum of transmitted power and the reflected power is equal to the total power sent into the ideal system. The reflected interferogram $I_{\mathrm{Refl.}}$ [see Eq. (4.1)] is complementary to the transmitted one. At $s = 0$ *e.g.*, all energy is transmitted to the detector and nothing reflected ($I_{\mathrm{Refl.}} = 0$). In general, a maximum of $I_{\mathrm{Refl.}}$ corresponds to a minimum of the transmitted interferogram and vice versa. If now the difference of the two interferograms [Eq. (4.1) and Eq. (3.2)] is taken, the mean value $I(\infty)$ drops out and only the oscillatory part is left

$$I(s) - I_{\mathrm{Refl}}(s) = 2\,\bar{I}(s) = 4\int_0^\infty I(\tilde{\nu})\cos(2\pi\tilde{\nu}s)\,d\tilde{\nu}.$$

But the experimental realization of this concept clearly suffers from the fact that one has to deal with two detectors which generally do not have equal responsivity etc. However, using also the interferogram reflected back to the source offers the possibility of comparing the spectra of two samples or two sources, *i.e.* the possibility of a double beam operation in Fourier spectroscopy (cf. the double beam interferometer by R. Hanel *et al.*[43]) and Section 6.2).

In addition to the beam splitter and the other optical elements, the drive of the movable mirror and the system to measure the path difference with high accuracy are essential parts of a Michelson interferometer. Modern instruments for the near- and middle-infrared (see Section 6) usually provide a rapid scan mechanism where the movable mirror is driven at a speed of round about 1 cm/sec. That means that it takes less then a second to scan an interferogram for an intermediate resolution, *e.g.* $T = 0.5$ sec for $v = 1$ cm/sec and $s_{\max} = 1$ cm. In this region, the movement of the mirror is generally monitored by interference fringes of a He-Ne-laser. In some instruments for the far infrared spectral region, the movable mirror is driven by a synchronous motor at a speed of round about $1 \cdot 10^{-3}$ cm/sec so that 10–20 min are needed to scan an interferogram. It should be noted that the signal-to-noise ratio is very poor in the far-infrared and relatively large integration times are needed. Therefore, step-motors are sometimes used for the mirror drive in this region. For these longer wavelengths, it is sufficient to use a Moiré-system to measure the path-difference instead of a He-Ne-laser (cf. Section 6).

### 4.2 Analogue Fourier Transform

The Fourier transform necessary to convert $I(s)$ into $I(\tilde{\nu})$ may be executed either in an analogue or in a digital way. In this section, we shall concentrate on the first possibility. One potential way is illustrated in Fig. 19: the movable mirror of the Michelson interferometer is moved according to a triangular wave function (Fig. 19 upper left). This can be done by means of an appropiate mechanical cam system. Problems arise only at the points of reversal, but usually these difficulties are bypassed by the apodization. During half a period the mirror moves with constant velocity, say $v_0$. Hence $s = 2v_0t$ and, for monochromatic radiation, there is an a.c. component in the signal recorded by the detector:

$$\bar{I}(t) = 2I_0\cos(2\pi\tilde{\nu}_0 s) = 2I_0\cos(2\pi[2\tilde{\nu}_0 v_0]\,t) \qquad (4.2)$$

with a frequency of $f_0 = 2\tilde{\nu}_0 v_0$. During a longer time, the interferogram is repeated periodically as the mirror moves back and forth (Fig. 19 lower left). The amplitude

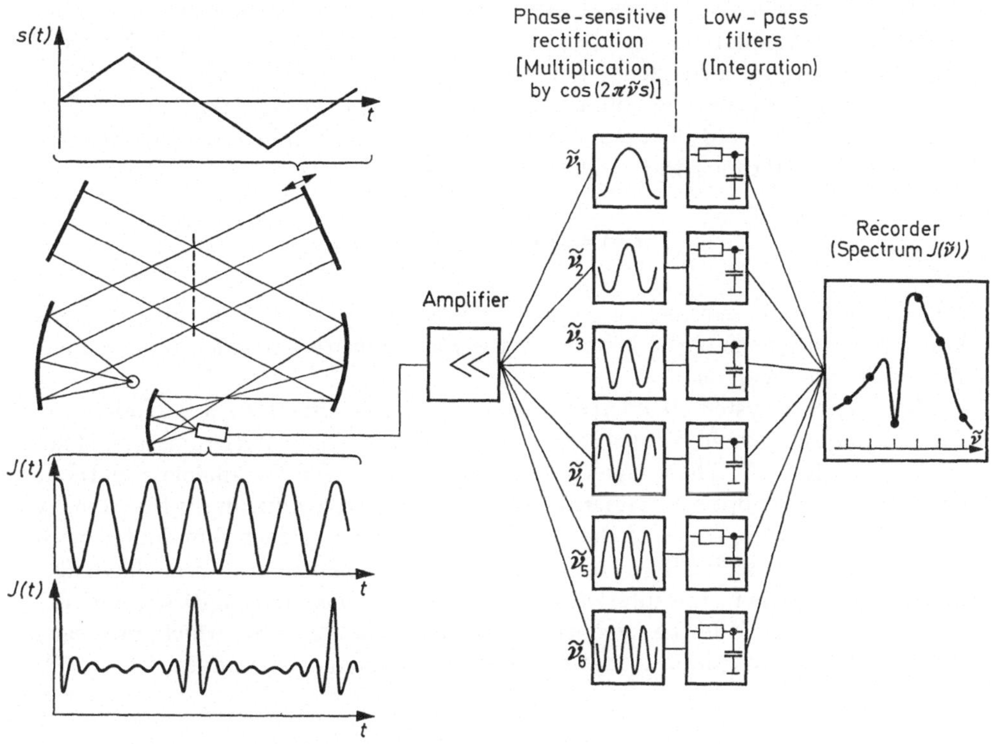


Fig. 19. Illustration of Fourier transform spectroscopy by means of an analogue computer

of the mirror motion is one half of the maximum path difference and therefore governs the resolution of this system. For a continuous spectrum, each spectral element of wave number $\tilde{\nu}$ is modulated with a frequency

$$f = 2\tilde{\nu}v_0 . \tag{4.3}$$

Because of this modulation effect, the system has been called "interference modulation"[22,26].

If we consider the periodically scanned interferogram as a function of time, *i.e.* as an a.c. signal, it is a superposition of all the components of frequency $f = 2\tilde{\nu}v_0$ according to the spectral distribution $I(\tilde{\nu})$. Fourier transform means in this case that the recorded signal must be analyzed with respect to its frequency components. Because of noise problems, the signal is first amplified in a broad-band amplifier, then the signal is transferred to a number of phase-sensitive synchronous rectifiers. In each of these the signal is multiplied by a sinusoidal reference signal of frequency $f_r$ and, by means of the following low-pass filter, the strength of the component of frequency $f = f_r$ is obtained from the amplified signal.

All the phase-sensitive rectifiers are tuned to different frequencies and their number should equal the number of spectral elements to be analyzed simultaneously. Now, each frequency $f_r$ is connected to a wave number $\tilde{\nu}$ in a unique way [see Eq. (4.3)] and the voltage at the low-pass filters is proportional to $I(\tilde{\nu})$. Consequently, a record of these voltages yields the spectrum $I(\tilde{\nu})$ (Fig. 19). The "computer" performing the Fourier transform

$$I(\tilde{\nu}) = \int \bar{I}(s) \cos(2\pi\tilde{\nu}s)\, ds$$

consists of the phase-sensitive rectifier and the low-pass filter in this case. The multiplication by $\cos(2\pi\tilde{\nu}s)$ is done by the phase-sensitive rectifier and the integration by the low-pass filter.

In practice, a system like that shown in Fig. 19 has been brought into operation with only one electronic channel and with a lamellar grating interferometer instead of the Michelson [26)]. There have been other proposals for the analogue method; one is the coherent optical Fourier transform [46)]. Here the difficulty is that a mask has to be cut in the shape of the interferogram, which is then illuminated with coherent laser light. Finally, it should be mentioned that hybrid forms have also been produced where the interferogram is scanned once and digitized, and the information is stored in the memory of a digital computer, after which, by means of an electronic control unit, it is passed through an analogue wave analyzer [48)].

### 4.3 Digital Fourier Transform

The method most frequently and widely used is the digital Fourier transform. A Fourier spectrometer employing this method is outlined schematically in Fig. 20. One of the mirrors is moved at constant speed $v$. The light in the interferometer is chopped at frequency $f_0$. In rapid-scan instruments, the chopper is missing. There, the signal is only modulated by the interference effects (Interference modulation, see Section 4.2). The signal at the detector is amplified and passed to the phase-sensitive rectifier. At the low-pass filter, the analogue signal proportional to the interferogram $I(s)$ is obtained. This is converted to the digital form, and the data are transferred to the computer where the Fourier transform is executed. The result, the spectrum $I(\tilde{\nu})$, may then be recorded or displayed on a screen after being converted back to an analogue signal (see Fig. 20).

Let us now consider some of the computational aspects. It has already been pointed out that for the digital way the Fourier transform integral is converted to a sum [see Eq. (3.8)]. The problem of aliasing in the spectrum [see Eq. (3.9) or (3.10) was also discussed. In the foregoing it was mostly assumed that the interferogram was measured as a double-sided one, *i.e.* from $-s_{\max}$ to $+s_{\max}$. Theoretically, the interferogram is symmetric about $s=0$. In practice, there may be small asymmetries due to misalignment of the interferometer. Nevertheless, often only a single-sided interferogram is recorded and subjected to a cos transform:

$$I_{\text{obs}}(\tilde{\nu}) = 2 \sum_{n=0}^{N} \bar{I}(n\,\Delta s) \cdot \cos(2\pi\tilde{\nu}n\,\Delta s) \qquad (4.4)$$

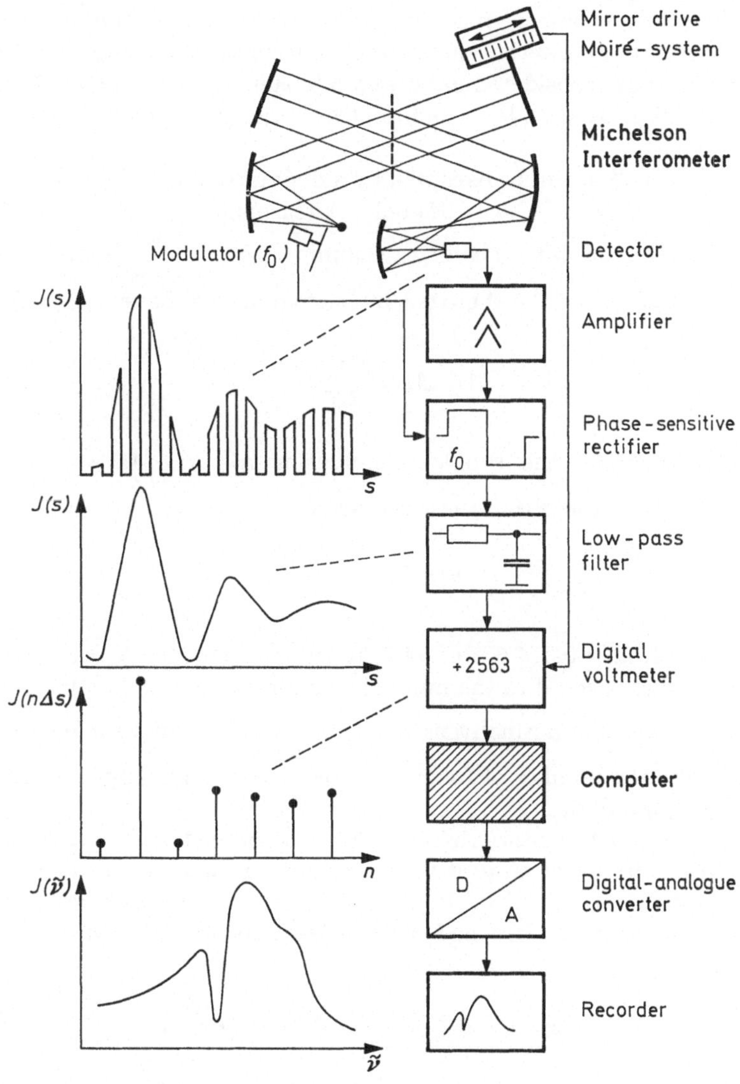


Fig. 20. Illustration of Fourier transform spectroscopy by means of a digital computer

where

$$\tilde{I}(n\Delta s) = S(n\Delta s) \cdot \bar{I}(n\Delta s) \cdot \Delta s \quad \text{for } n \geqq 1 \tag{4.5}$$

and

$$\tilde{I}(0) = \frac{1}{2} S(0) \cdot \bar{I}(0) \cdot \Delta s \qquad \text{for } n = 0$$

is the oscillatory part of the interferogram already multiplied by the apodization function $S$ and the sampling interval $\Delta s$. The summation [see Eq. (4.4)] requires that the first point ($n = 0$) coincides exactly with the great maximum of $I(s)$ at zero path difference, otherwise phase errors will occur. These, and also errors due to misalignment of the interferometer, their effects on the computed spectrum, and their removal are considered in Section 5.3. Here, we concentrate first on the main concepts. We assume ideal conditions and disregard possible errors for the present.

The sum in Eq. (4.4) can be executed by a digital computer for a finite number of frequencies $\tilde{\nu} = m\Delta\tilde{\nu}$ ($0 \leq m \leq M$) only. A reasonable choice of $\Delta\tilde{\nu}$ is to make it equal to the resolved difference in wave numbers $\Delta\tilde{\nu} = \frac{1}{2s_{max}}$ [without apodization, see Eq. (2.21)]. For the digitized interferogram, we have $s_{max} = N\Delta s$ and consequently

$$\Delta\tilde{\nu} \cdot \Delta s = \frac{1}{2N} \,. \tag{4.6}$$

On the other hand, the problem of aliasing requires $\Delta s = \frac{1}{2\tilde{\nu}_{max}}$, where $\tilde{\nu}_{max} = M \cdot \Delta\tilde{\nu}$ in this case. From these relations

$$\Delta\tilde{\nu} \cdot \Delta s = \frac{1}{2M} \tag{4.7}$$

is obtained, and from Eqs. (4.6) and (4.7) it follows that $M = N$, *i.e.* the number of frequency points is equal to the number of interferogram points for our choice of $\Delta\tilde{\nu}$. As apodization is normally used, it may be sufficient to make $\Delta\tilde{\nu} = \frac{1}{s_{max}}$ and, accordingly, the number of frequency points $M$ will be only one half of that of the interferogram data $N$.

Now let us assume the interferogram is already completely measured and is fed directly to the computer or on to it via paper tape or via cards. A diagram of the computation of the spectrum is shown in Fig. 21. The interferogram data are stored in the computer memory, and the Fourier transform [substituting Eq. (4.6) into Eq. (4.4)]

$$I_{obs}(m\Delta\tilde{\nu}) = 2\sum_{n=0}^{N} \bar{I}(n\Delta s)\cos\left(\pi\frac{mn}{N}\right) \tag{4.8}$$

is performed successively for all the frequencies $m\Delta\tilde{\nu}$ wanted. This method can also be applied to double-sided interferograms and for performing sin and cos transforms (see Section 5.3) and not only the cos transform, as in our example. The computer time needed in this case is proportional to $N^2$, since there are $N^2$ operations consisting of one multiplication by a phase factor $\cos\left(\pi\frac{m \cdot n}{N}\right)$ (see Fig. 21) and one addition. Much computer time can be saved if the Cooley-Tukey algorithm is used instead of the conventional integration.

The Cooley-Tukey algorithm is a rather involved mathematical formalism [49, 50] and cannot be explained here in full detail. A short derivation with an illustration of the method for $N = 4$ is given in Appdx 2. Here we use only the result developed there, that the required computer time is proportional to $N \cdot {}_2\log 2N$ because this is the number of operations to be executed in the course of a compu-

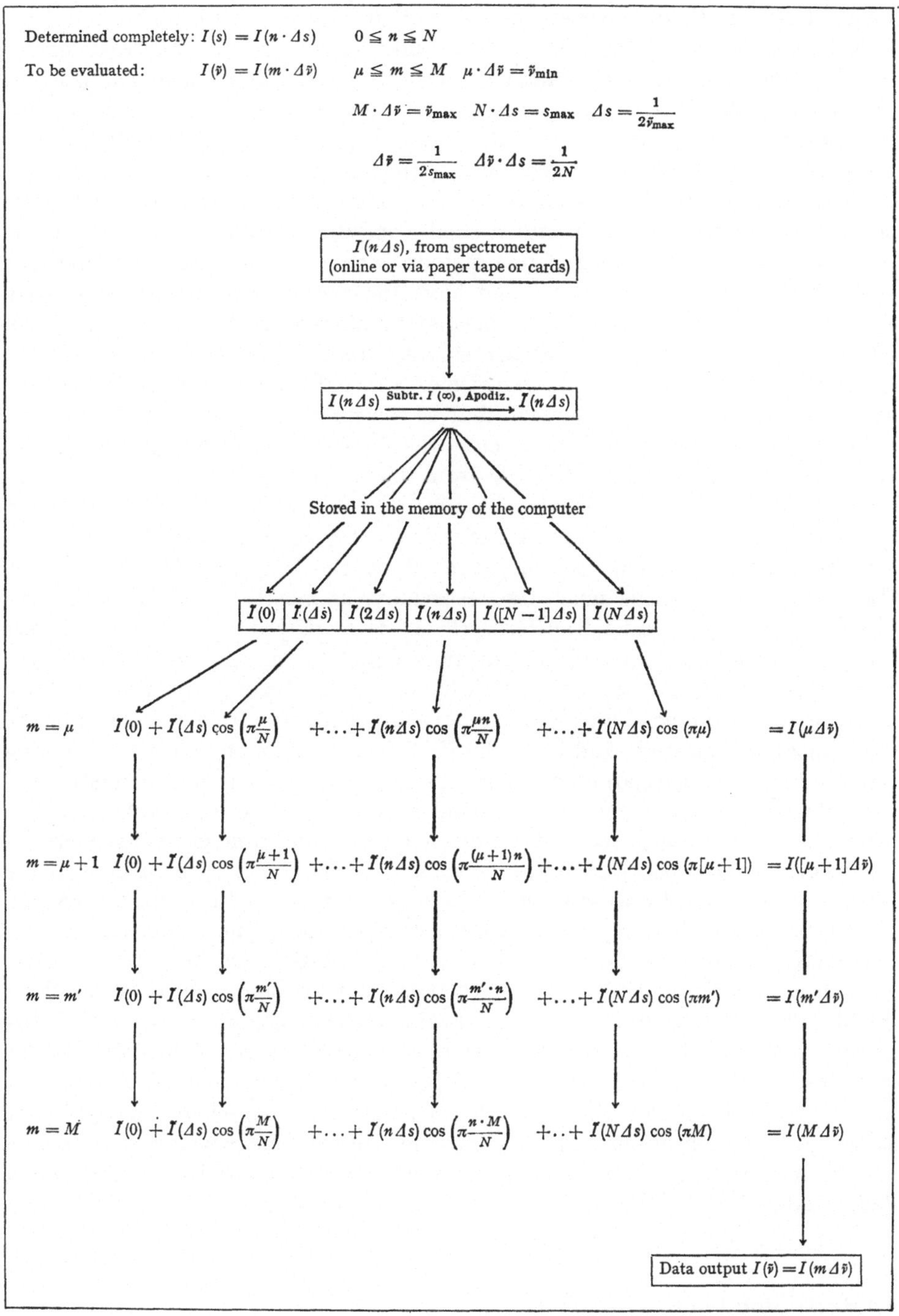


Fig. 21. Computation of the spectrum $I(\tilde{\nu})$ in the usual digital way, when the interferogram $I(s)$ has been determined completely

tation. In this method, however, the calculation is done with complex numbers, and the time necessary for one operation will therefore greater than that for one operation in the conventional method. The extra time is more than compensated for by having the factor ${}_2\log 2N$ instead of $N$ for large $N$. Consequently, for the processing of a large amount of data (large $N$), the Cooley-Tukey algorithm is more economical than the conventional method in use of computer time.

The method described so far is now in common use in a number of Fourier spectrometers most of which have a small computer attached to it (see Section 6.4). The main disadvantage of this method is that it is necessary to determine the complete interferogram first and have the spectrum after that evaluated. If the interferogram is recorded by means of a slow-scan FIR instrument and if electronic data processing is available only from a computation center, the experimeter will to have to wait some time, eventually a day, until he can tell the success of his experiment. We recall that he can judge the quality of his measurement only from the interferogram before the Fourier transform is performed and that it is difficult to gather all details from the interferogram. And even if an online computer or time-sharing facility are used in connection with a slow-scan interferometer, the experimenter has to wait for the spectrum until the scan of the interferogram is finished. The time needed by the computer for the Fourier-transform of the data is much less, especially, as mostly the Cooley-Tukey algorithm or fast Fourier-transform (FFT) is employed instead of the direct integration method discussed here in more detail for merely tutorial purposes. For rapid-scan instruments on the other hand, the disadvantage mentioned above does not count so much because the spectra are available within seconds after the experiment was started. And here, the FFT (and also the more time consuming direct integration method) offers some advantages. Especially in the middle- and near-infrared spectral region, it is difficult to avoid phase errors (cf. Section **5.3**) originating from misalignment of the interferometer and from errors in determining accurately zero path-difference. In rapid-scan instruments, distortions of the interferogram may also arise from the limited bandwith of the amplifier system. As will be discribed later (Section **5.3**) the computational method discussed here is suitable for correcting errors as it can handle single-sided as well as double-sided interferograms. And the most economical way is to take a rather short double-sided interferogram in order to evaluate the errors and to scan the interferogram only single-sided up to full resolution which then will be corrected for the errors already determined. For all these reasons, commercial instruments (see Section 6.4) employ this method of computation, especially the FFT. For some special applications, *e.g.* asymmetric Fourier spectroscopy (see Section 4.7), the interferogram is no longer symmetric about $s=0$ and has to be recorded double-sided.

## 4.4 Real-Time Fourier Analysis

The arguments in the last section have shown that there is a need for a different computational method for slow-scan FIR instruments. There it would be a great advantage to have the spectrum computed at the same time as the interferogram is being recorded. This advantage is provided by the so-called real-time Fourier analysis [51–53], a method which is explained below. It can be applied in a reasonable way only to one-sided interferograms ($0 \leqslant s \leqslant s_{max}$). For the application of this method therefore, a special effort should be made to obtain a symmetric interferogram and to obtain the first sample point exactly at $s=0$. There is only a limited possibility of correcting linear phase errors in this case (see Section **5.3**). Here again, let us assume the ideal case where $I(n\Delta s)$ has been determined for $N$ sample points $s=N\,\Delta s$ ($0 \leqslant n \leqslant N$), starting at $s=0$ ($n=0$). As in the previous section, our choice for $\Delta\tilde{\nu}$ is the ideal value $\Delta\tilde{\nu}=\frac{1}{2s_{max}}=\frac{1}{2N\Delta s}$. Then, $I(\tilde{\nu})$ is to be evaluated for $M=N$ values of $\tilde{\nu}=m\Delta\tilde{\nu}$ ($0 \leqslant \tilde{\nu} \leqslant \tilde{\nu}_{max}$ or $0 \leqslant m \leqslant M$). As mentioned already, for practical cases $\Delta\tilde{\nu}=\frac{1}{s_{max}}$ may be chosen when apodization is used. In practice, the number of frequency points is often further reduced by starting at the low-frequency end $\tilde{\nu}_{min}=\mu\Delta\tilde{\nu}$ instead of at $\tilde{\nu}=0\,(m=0)$; then the actual number of frequency points is $M-\mu=\frac{1}{2}N-\mu$. For our considerations, however, we again assume the maximum number $M=N$, and Eqs. (4.4) to (4.8) hold for the computation of $I(\tilde{\nu})$. In other words, we have the same conditions, the same assumptions and therefore the same mathematical problem to compute $I(m\Delta\tilde{\nu})$ from $I(n\Delta s)$ as in Section 4.3. The only difference is in the method or the way used for the actual execution of the computation.

This method, real-time Fourier analysis, is demonstrated schematically in Fig. 22. The principle is that the computer processes the interferogram data as soon as they have been measured with the interferometer. The procedure starts with the main maximum of the interferogram at $s=0$. In the first step, the value $I(0)$ is transferred from the spectrometer to the computer. There the mean value $I(\infty)$ is subtracted and multiplied by the apodization function to give the value $\tilde{I}(0)$. This is the final form of the interferogram data as needed for the computation [see Eq. (4.8)]. Now, not the interferogram data but the frequency data are stored in the memory of the computer in this method. Therefore, the locations of the memory are assigned to the numbers $m$ of the frequency points and not to the numbers $n$ of the interferogram points, as is the case when the computation is performed after the interferogram has been completely recorded. To continue the first step of the real-time method, the prepared interferogram value $\tilde{I}(0)$ is multiplied by the phase factors $\cos\left(\pi\frac{m\cdot 0}{N}\right)$ and then stored in the memory at location no. $m$. In the second step, $I(1\Delta s)$ is obtained from the interferometer and is converted to $\tilde{I}(1\Delta s)$. Then, this value is multiplied by the phase factors $\cos\left(\pi\frac{m\cdot 1}{N}\right)$ and the results are added to the contents of the computer memory at location no. $m$. Whenever an interferogram point $I(n\Delta s)$ is measured, it is converted immediately to $\tilde{I}(n\Delta s)$. After multiplication of the phase factors $\cos\left(\pi\frac{m\cdot n}{N}\right)$ the products are added to the contents in number $m$. The last step in this sequence is the $(N+1)$th with $s=N\Delta s$. When it is finished, the contents of the computer memory are the values $I(m\Delta\tilde{\nu})$ for the different frequencies $\tilde{\nu}=m\Delta\tilde{\nu}$, and the Fourier transform is completed.

To be evaluated: $I(\tilde{\nu}) = I(m \cdot \Delta\tilde{\nu})$    $\mu \le m \le M$    $\mu \cdot \Delta\tilde{\nu} = \tilde{\nu}_{min}$    $M \cdot \Delta\tilde{\nu} = \tilde{\nu}_{max}$

Being determined: $I(s) = I(n \cdot \Delta s)$    $0 \le n \le N$

$$N \cdot \Delta s = s_{max} \quad \Delta s = \frac{1}{2\tilde{\nu}_{max}} \quad \Delta\tilde{\nu} = \frac{1}{2 s_{max}} \quad \Delta s \cdot \Delta\tilde{\nu} = \frac{1}{2N}$$

1)  $s = 0$  From spectrometer: $I(0) \xrightarrow[\text{Apod.}]{\text{Subtr. } I(\infty)} I(0)$

Multiplied by $\cos\left(\dfrac{\pi m \cdot 0}{N}\right)$ and stored in the memory at loc. no. $m$

$m \rightarrow$ | $I(0)$  $\mu$ | ..... | $I(0)$  $m'$ | ..... | $I(0)$  $M$ |

2) $s = \Delta s$  From spectrometer: $I(\Delta s) \xrightarrow[\text{Apod.}]{\text{Subtr. } I(\infty)} I(\Delta s)$

Multiplied by $\cos\left(\dfrac{\pi m}{N}\right)$ and added to the contents at loc. no. $m$

| $I(0) + I(\Delta s) \cos\left(\dfrac{\pi\mu}{N}\right)$ | ..... | $I(0) + I(\Delta s) \cos\left(\dfrac{\pi m'}{N}\right)$ | ..... | $I(0) + I(\Delta s) \cos\left(\dfrac{\pi M}{N}\right)$ |

$n + 1)\, s = n\Delta s$  From spectrometer: $I(n\Delta s) \xrightarrow[\text{Apod.}]{\text{Subtr. } I(\infty)} I(n\Delta s)$

Multiplied by $\cos\left(\dfrac{\pi m n}{N}\right)$ and added to the contents at loc. no. $m$

| $\sum I(n\Delta s) \cos\left(\dfrac{\pi\mu n}{N}\right)$ | ..... | $\sum I(n\Delta s) \cos\left(\dfrac{\pi m' n}{N}\right)$ | ..... | $\sum I(n\Delta s) \cos\left(\dfrac{\pi M n}{N}\right)$ |

$N + 1)\, s = N\Delta s$  From Spectrometer: $I(N\Delta s) \xrightarrow[\text{Apod.}]{\text{Subtr. } I(\infty)} I(N\Delta s)$

Multiplied by $\cos\left(\dfrac{\pi m N}{N}\right)$ and added to the contents at loc. no. $m$

| $I(\mu\Delta\tilde{\nu})$ | ..... | $I(m'\Delta\tilde{\nu})$ | ..... | $I(M\Delta\tilde{\nu})$ |

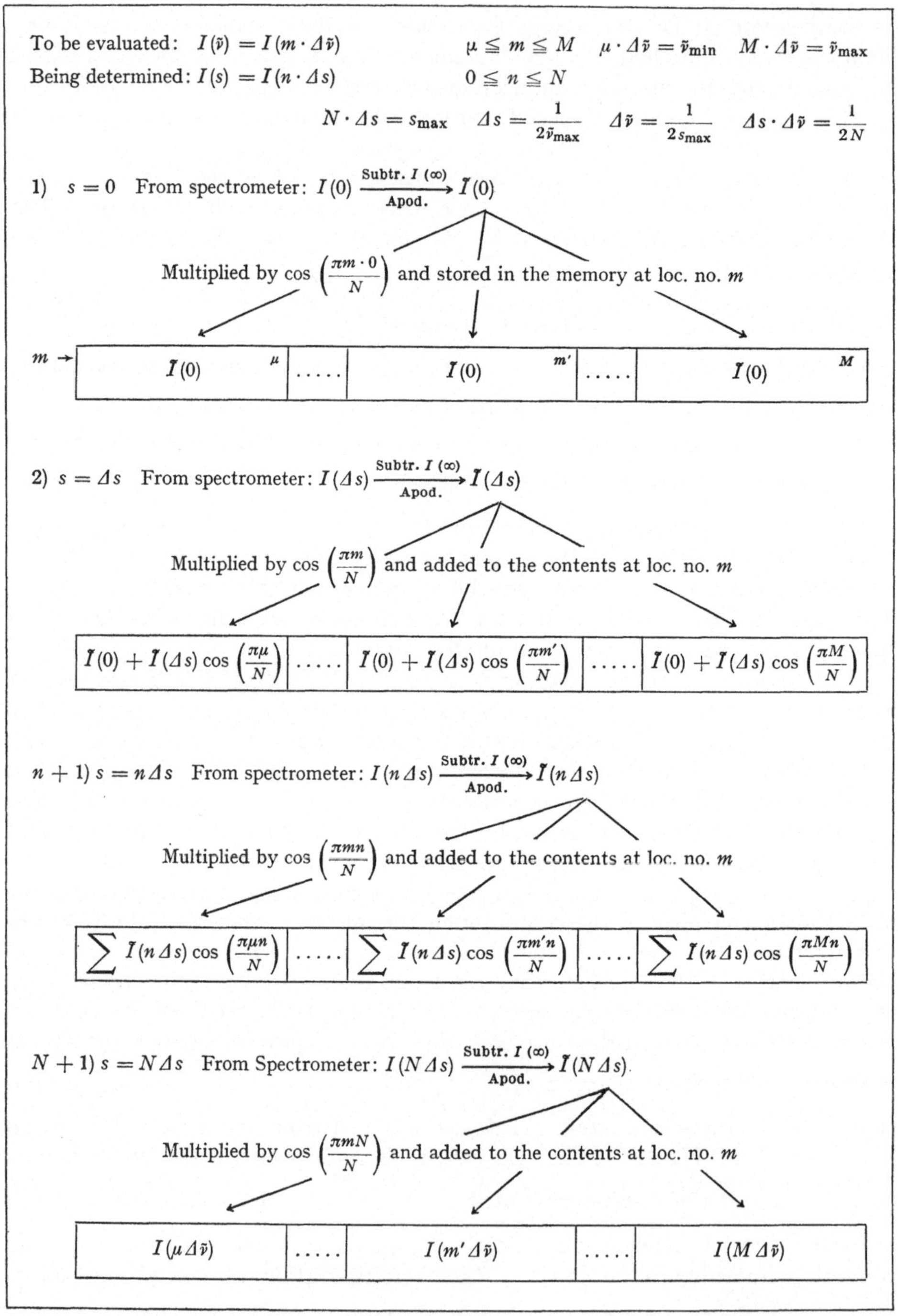

Fig. 22. Computation of the spectrum $I(\tilde{\nu})$ in the real-time method, when the interferogram $I(s)$ is being determined

As mentioned above, the major advantage of this method is that the experimentalist is able to watch the spectrum $I(\tilde{\nu})$ while the experiment is under way. This is achieved by means of a fast electronic reading of the contents of the computer memory after each step and by displaying it on the screen of an oscilloscope. It is obvious that this method is feasible only if a small Fourier transform computer is attached to the Fourier spectrometer or a large central computer can be used "on line". It should be noted that the computation time for each step is usually small in comparison to the time needed to obtain one interferogram point, due to the large time constants necessary in far-infrared spectroscopy. The spectrum displayed on the screen is dependent on the path difference reached so far in the interferogram. Sections 2 and 3 outlined the way in which the resolution is inversely proportional to the maximum path difference and the interferogram pattern close to $s=0$ is due to the gross features of the spectrum and those for larger $s$ are due to the finer details (see Figs. 11 and 12). With the real-time method, it is possible to see and study directly what was derived theoretically in the previous sections. An example is shown in Fig. 23. A number of curves $I(\tilde{\nu})$ show the spectrum on the screen at various path differences, as indicated in the interferogram. At small path differences, only a coarse outline of the spectrum is seen. With increasing path difference, more and more details of the spectrum appear. In our example, this is first the region of high reflection due to lattice vibrations in NaCl (reststrahlen band), then the secondary maxima in the reflectivity, and

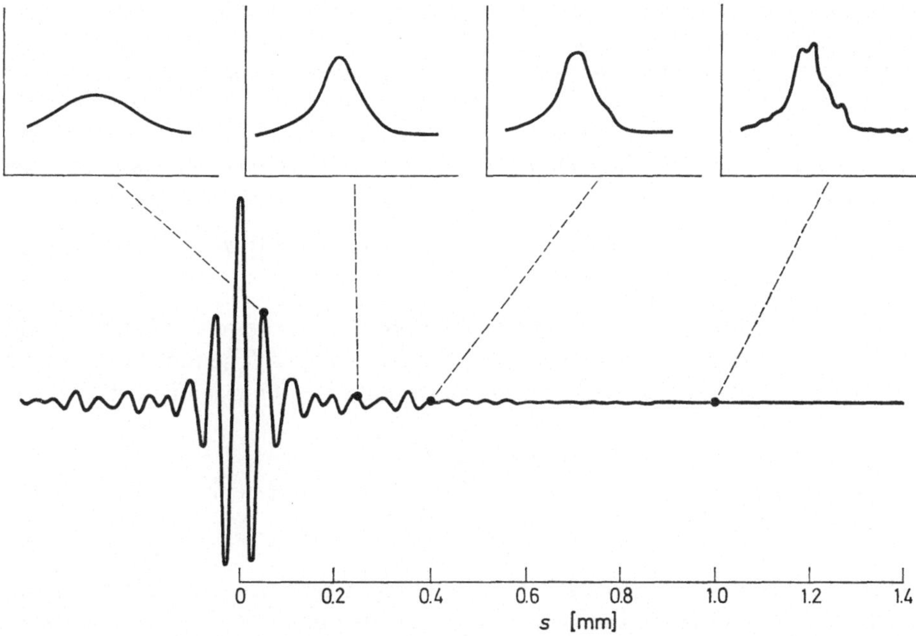


Fig. 23. An example of the real-time method: spectrum (50 to 400 $cm^{-1}$) with one reflection at an NaCl single crystal. The curves $I(\tilde{\nu})$ show the spectra displayed on the oscilloscope screen for the different paths travelled by the movable mirror indicated in the interferogram. These data were obtained with a Polytec FIR 30 Fourier spectrometer

finally the channel spectra, due to interference from zigzag reflections in a plane-parallel window. This evolution of the spectrum on the oscilloscope screen during the experiment is the useful tool for checking the success of the experiment in the case of slow-scan long-time far infrared measurements.

### 4.5 Phase Modulation

For many applications, there may be some advantage in employing phase modulation [54,55] instead of the usual amplitude modulation. In the latter technique the path of the radiation from the source to the detector is blocked and opened periodically by a chopper (cf. Fig. 20 and Section 4.3). For phase modulation, the chopper is removed from the spectrometer and the fixed mirror of the Michelson interferometer is moved back and forth about its mean position with a certain frequency. In contrast to the interference modulation (see Section 4.2), the amplitude of the mirror motion is small, being a quarter of the wavelength of the light. For the analogue Fourier transform or interference modulation, the amplitude of the mirror has to have many wavelengths in order to achieve a reasonable resolution $R = \tilde{\nu} \cdot s_{\max} = \frac{s_{\max}}{\lambda}$ [see Eqs. (2.23) and (2.27)]. Here, we are still dealing with the case where the interferogram is scanned once as a function of path difference $s$ or of the position of the movable mirror, which is driven at constant speed. The Fourier transform is executed in a digital way. The motion, or rather oscillation of the fixed mirror is to ensure that the signal at the detector is modulated to permit the use of the usual lock-in technique with narrow-band amplifier and phase-sensitive rectifier (Fig. 24).

Though a sinusoidal motion of the mirror is easier to carry out and is most often used, let us assume for simplicity that it moves according to a square-wave function. Then the intensity at the detector, the interferogram, is a function of the path difference $s$ and of the time $t$ [for the case of a continuous spectrum, see Eq. (3.2)]:

$$I_{\varphi}(s,t) = 2 \int_0^\infty I(\tilde{\nu}) \{1 + \cos(2\pi\tilde{\nu}[s + \sigma(t)])\} d\tilde{\nu} \qquad (4.9)$$

where $s$ is path difference resulting from the position of the movable mirror and $\sigma(t)$ that resulting from the oscillation of the fixed mirror. The index $\varphi$ indicates the phase modulation. For the assumed square-wave motion (see Fig. 24):

$$\sigma(t) = \begin{cases} +\sigma_0 & 0 < t < \frac{1}{2} T_0 \\ -\sigma_0 & \frac{1}{2} T_0 < t < T_0 \end{cases} \quad \text{and} \quad \sigma(t + T_0) = \sigma(t)\,. \qquad (4.10)$$

In Eq. (4.9), $s$ is not expressed as a function of time since the motion of the movable mirror and hence the variation of $s$ in time must be slow in comparison to the modulation $\sigma(t)$. From Eq. 4.9 we gather further that $2\pi\tilde{\nu}\sigma(t)$ in the argument of the cosine function is an oscillating phase, so this procedure is called phase modulation. In the narrow-band amplifier, only the a.c. component of $I_{\varphi}(s,t)$ with frequency $f_0 = 1/T_0$ is amplified to which frequency the amplifier is tuned. All

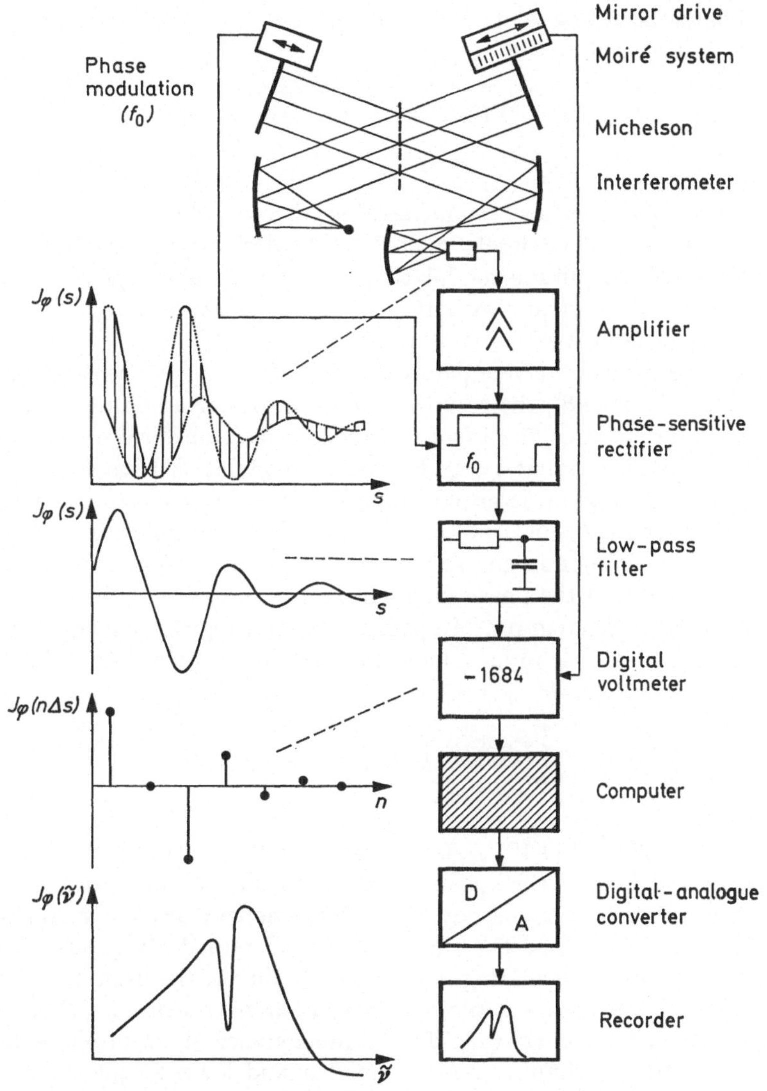


Fig. 24. Illustration of Fourier transform spectroscopy when phase modulation is employed instead of the usual amplitude modulation

other a.c. components with frequencies $nf_0$ and also the d.c. component are suppressed. It should be noted that suppression of the d.c. component means we get rid of $I(\infty)$, which had to be subtracted from $I(s)$. This helps to avoid errors introduced by an erroneous determination of $I(\infty)$. The details of the Fourier analysis of $I_\varphi(s,t)$ are given in Appdx 3. For the output signal of the amplifier, the result is

$$\bar{I}_\varphi(s,t) \sim \{2 \int_0^\infty I(\tilde{\nu}) \sin(2\pi\tilde{\nu}s) \sin(2\pi\tilde{\nu}\sigma_0)\, d\tilde{\nu}\} \sin(2\pi f_0 t)\,. \tag{4.11}$$

Accordingly, the analogue signal obtained at the low-pass filter is proportional to

$$\bar{I}_{\varphi}(s) = -2\int_{0}^{\infty} I(\tilde{\nu}) \sin(2\pi\tilde{\nu}\sigma_0) \sin(2\pi\tilde{\nu}s)\, d\tilde{\nu} \,. \tag{4.12}$$

Here the negative sign has been included for theoretical reasons (see Appdx. 3). The interferogram $\bar{I}_{\varphi}(s)$ is closely related to the derivative of the interferogram $\bar{I}(s)$ obtained with amplitude modulation (see Fig. 24 and Appdx 3). For the computation of the spectrum with the computer, the interferogram has to be digitized in the usual way.

If the Fourier transform were performed as in all the previous cases according to Eqs. (3.6), (4.4) or (4.8), the result would be zero. For amplitude modulation, the interferogram [see Eq. (3.2)] is a so-called cosine transform of the spectrum $I(\tilde{\nu})$. It is symmetrical about $s=0$. For phase modulation [(see Eq. (4.12) and Fig. 24)], the interferogram is an odd function of $s$, *i.e.* $\bar{I}_{\varphi}(-s) = -\bar{I}_{\varphi}(s)$. This is the so-called sine transform of the spectrum $I(\tilde{\nu})$ multiplied by $\sin(2\pi\tilde{\nu}\sigma_0)$. For these reasons, we obtain the spectrum by a sine transform. In order to demonstrate the effects of the phase modulation in a simple way, let us neglect all effects of a finite interferogram, of apodization, and digitization for a moment, and consider the ideal case of an infinite interferogram. Then we obtain

$$I_{\varphi}(\tilde{\nu}) = -\int_{-\infty}^{\infty} \bar{I}_{\varphi}(s) \sin(2\pi\tilde{\nu}s)\, ds = I(\tilde{\nu}) \sin(2\pi\tilde{\nu}\sigma_0) \,. \tag{4.13}$$

From Eqs. (4.12) and (4.13) and also from Fig. 24, we see the main effects of phase modulation. The interferogram is asymmetric and is similar to the observed signal in conventional spectroscopy when the magnetic field is modulated for Zeeman effect investigations. The effect can be understood if we consider that in our case, because of the oscillating mirror, we obtain not the usual interferogram but something related to its derivative. The Fourier transform yields the true spectrum multiplied by $\sin(2\pi\tilde{\nu}\sigma_0)$. The suitable choice of $\sigma_0$ provides the possibility to shift the maximum of $I(\tilde{\nu})$ to higher and lower frequencies. In other words, this factor can be used as a kind of filter (see Fig. 25). For $\sigma_0 = 10$, 25, or 100 μm, the first maximum of $\sin(2\pi\tilde{\nu}\sigma_0)$ would be at $\tilde{\nu} = 250$, 100, or 25 $cm^{-1}$ respectively. All the effects of finite interferogram etc. are the same as discussed earlier and need not be repeated in this context.

In practice, a sinusoidal motion of the oscillating mirror is to be preferred to the square-wave motion, and in most applications, the sinusoidal motion is used. The effects of phase modulation are the same in this case, the only difference being that the modulation factor in Eqs. (4.11) to (4.13) will be the first Bessel function $J_1(2\pi\tilde{\nu}\sigma_0)$ instead of $\sin(2\pi\tilde{\nu}\sigma_0)$. The first maximum of the Bessel function is at $\tilde{\nu} = \frac{0.29}{\sigma_0}$ instead of $\tilde{\nu} = \frac{0.25}{\sigma_0}$ for the sine function. An example in Fig. 25 demonstrates how sinusoidal phase modulation is a very useful tool and filter for spectroscopy in the extreme far-infrared [56].

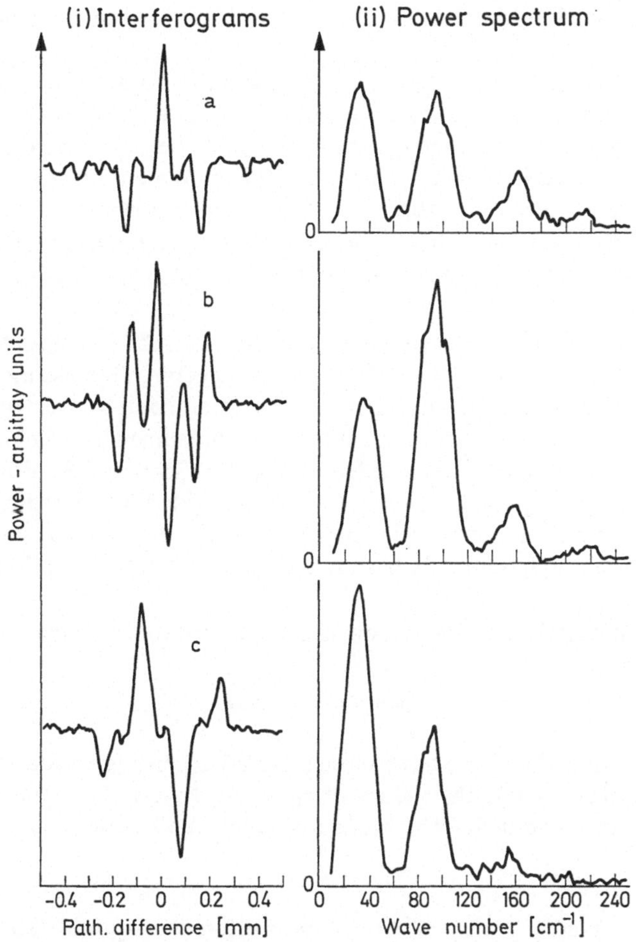


Fig. 25. Interferograms recorded using (a) amplitude modulation; (b) sinusoidal phase modulation, vibration amplitude 30 $\mu$m; and (c) sinusoidal phase modulation, vibration amplitude 100 $\mu$m. The corresponding spectra are shown for comparison. All data were to taken from Ref. [56)]

## 4.6 Examples

Summarizing the results of our discussion of the practice of Fourier transform spectroscopy, we start with the presumption that the equipment for most routine spectroscopic investigations consists of a Fourier spectrometer with a Michelson interferometer and a digital computer. In other words, the advantages of the lamellar grating used as a two-beam interferometer, and of phase modulation, for example, have been utilized only for certain special applications in the extreme far-infrared. All commercial Fourier spectrometers are available with a computer attached, which in most cases not only performs the Fourier transform but is also programmed to control the instrument. Commercial instruments have a remote switch for the selection of the different spectral ranges, and the filters and beams splitters are changed automatically. What then is left for the spectroscopist to do and to think about?

Obviously, the experimentalist has to decide on the basis of the problem to be studied whether he can extract more useful information by reflection or by transmission measurements. Another question to which he has to find an answer is whether the investigation should be performed at room temperature, at low or even at elevated temperature. But apart from these questions, which are closely related to the physical or chemical or biological problem under investigation, there are still a few problems connected with Fourier transform spectroscopy:

1. First, the spectroscopist has to decide whether to record a one-sided or a two-sided interferogram. The latter will be advantageous in bypassing phase errors and noise problems. As already mentioned, a good compromise with respect to phase errors is to start scanning the interferogram somewhere before zero path difference and to extend it single-sided to $s_{max}$ (cf. Section 5.3). When real-time Fourier analysis is applied, however, only a one-sided interferogram can be recorded. In many of the commercial instruments, the decision about one- ore two-sided interferogram has already been made by the manufacturer (see Section 6).

2. The spectrocopist has to decide on the most suitable spectral range for his investigation:

$$\tilde{\nu}_{min} \leqq \tilde{\nu} \leqq \tilde{\nu}_{max} .$$

   On the one hand, the filters have to be selected according to the value of $\tilde{\nu}_{max}$, and on the other hand, the values of $\tilde{\nu}_{min}$ and $\tilde{\nu}_{max}$ have to be fed into the computer in order to define the range for which the Fourier transform is to be executed.

3. For optical investigations, a certain resolution is required. In the case of Fourier spectroscopy, this is expressed as the smallest clearly resolved difference in wave numbers:

$$\Delta\tilde{\nu} = \frac{1}{s_{max}} .$$

   From the required value of $\Delta\tilde{\nu}$, the maximum path difference $s_{max}$ up to which the interferogram must be recorded is obtained (cf. Table 1).

4. When decisions have been made about the values of $\tilde{\nu}_{min}$, $\tilde{\nu}_{max}$, and $\Delta\tilde{\nu}$ or $s_{max}$, then the sampling interval $\Delta s$ has to be chosen such that

$$\Delta s \leqq \frac{1}{2\tilde{\nu}_{max}} .$$

   While for $\tilde{\nu}_{min}$, $\tilde{\nu}_{max}$ and $s_{max}$ any value can be selected within certain limits continuously, there is mostly only a set of discrete values available for $\Delta s$. This originates from the Moiré or laser system for determining the path difference.

5. The spectroscopist should also pay attention to the question whether the sampling interval $\Delta s$, the speed $v$ of the movable mirror and the time constant $\tau$ of the electronic data recording system are in a proper relation with each other. It is well known (see also Section 5.4) that a low pass filter with time constant $\tau$ has the complex transfer function (amplitude and phase!)

$$[1 + i(2\pi f\tau)]^{-1}$$

and only signals with frequencies for which $f\tau < 1$ are passed by it without considerable attenuation. Now, the maximum wave number $\nu_{max}$ in the spectrum corresponds to a maximum frequency in the interferogram

$$f_{max} = 2\,v\,\tilde{\nu}_{max}\,.$$

An acceptable choice for $\tau$ is the following (cf. Section 5.4)

$$f_{max} \cdot \tau \approx 0{,}3$$

or

$$\tau \approx \frac{0{,}3}{f_{max}} = \frac{0{,}3}{2v\tilde{\nu}_{max}} = 0{,}3\,\frac{\Delta s}{v}$$

That means that the time constant $\tau$ has to be chosen less than one third of the time needed to scan just one sampling intervall. Otherwise the interferogram will be distorted by the electronic system in an untolerable way.

All these are the parameters for a measurement with a Fourier spectrometer.

It should be mentioned that in some cases the input data for the computer are not $\tilde{\nu}_{max}$ and $s_{max}$ but $M' = \frac{\tilde{\nu}_{max} - \tilde{\nu}_{min}}{\Delta\tilde{\nu}}$ and $N = \frac{s_{max}}{\Delta s}$, *i.e.* the number of frequency and interferogram points, respectively (cf. Table 1). Sometimes it is advisable to evaluate $I(\tilde{\nu})$ for more than one frequency point per resolution with $\Delta\tilde{\nu}$. Then, the number of spectrum points is a multiple of $M'$. These numbers have to be kept within the limits of the accessible capacity of the computer memory. For conventional integration, as already mentioned, the interferogram is stored in the memory; therefore, the capacity of a small computer limits the number $N$, *i.e.* $s_{max}$, and the resolution. For real-time Fourier analysis, we recall that the frequency points are assigned to the locations of the memory. The computer capacity limits the number of frequency points in this case. If $M'$ exceeds the available capacity, it is better to decrease $M'$, not by increasing $\Delta\tilde{\nu}$, but by dividing $\tilde{\nu}_{max} - \tilde{\nu}_{min}$ into two or more parts for each of which the calculation has to be performed separately. Before starting the actual measurement, it is advisable to check the alignment of the interferometer by inspecting the symmetry of the interferogram at $s = 0$. Especially for real-time Fourier analysis, it is also advisable to see that one digital point is close to $s = 0$.

A great variety of problems can be studied in the far-infrared by means of Fourier transform spectroscopy: vibrations of molecules and crystal lattices, rotation of molecules, electronic transitions in paramagnetic ions and semicon-

Table 1: Examples of properly chosen parameters in Fourier spectroscopy

| Spectral range $\tilde{\nu}_{min}$—$\tilde{\nu}_{max}$ (cm$^{-1}$) | Sampling interval $\Delta s$[1]) ($\mu m$) | Resolution $\Delta\tilde{\nu}$ (cm$^{-1}$) | Maximum path difference $s_{max}$[2]) (mm) | Number of interferogram points $N = s_{max}/\Delta s$ | Number of spectrum points $M = 2\frac{\tilde{\nu}_{max}-\tilde{\nu}_{min}}{\Delta\tilde{\nu}}$ [3]) |
|---|---|---|---|---|---|
| 800—10000 | 0.3164[4]) | 0.06 | 167 | 528000 | 307000 |
| 800—10000 | 0.3164 | 0.10 | 100 | 316000 | 184000 |
| 800—10000 | 0.3164 | 0.25 | 40 | 126000 | 74000 |
| 800—10000 | 0.3164 | 0.5 | 20 | 63000 | 37000 |
| 400— 4000 | 1.2656[4]) | 0.06 | 167 | 132000 | 120000 |
| 400— 4000 | 1.2656 | 0.10 | 100 | 79000 | 72000 |
| 400— 4000 | 1.2656 | 0.25 | 40 | 32000 | 28800 |
| 400— 4000 | 1.2656 | 0.5 | 20 | 16000 | 14400 |
| 400— 4000 | 1.2656 | 2.0 | 5 | 4000 | 3600 |
| 100— 1000 | 5.0 | 0.1 | 100 | 20000 | 18000 |
| 100— 1000 | 5.0 | 0.5 | 20 | 4000 | 3600 |
| 100— 1000 | 5.0 | 2.0 | 5 | 1000 | 900 |
| 50— 400 | 10.0 | 0.1 | 100 | 10000 | 7000 |
| 50— 400 | 10.0 | 0.5 | 20 | 2000 | 1400 |
| 50— 400 | 10.0 | 2.0 | 5 | 500 | 350 |
| 10— 100 | 40.0 | 0.1 | 100 | 2500 | 1800 |
| 10— 100 | 40.0 | 0.5 | 20 | 500 | 360 |
| 10— 100 | 40.0 | 2.0 | 5 | 125 | 90 |

[1]) Theoretically: $\Delta s = \frac{1}{2\tilde{\nu}_{max}}$; in practice: the instrumentally available value close to but not larger than $\frac{1}{2\tilde{\nu}_{max}}$

[2]) $s_{max} = \frac{1}{\Delta\tilde{\nu}}$ (for apodized interferogram).

[3]) Two spectrum points per resolution width $\Delta\tilde{\nu}$; theoretically $M = N$, if $\Delta s = \frac{1}{2\tilde{\nu}_{max}}$ and if $\tilde{\nu}_{min} = 0$.

[4]) $0.5 \times 0.6328$ $\mu$m (He-Ne-laser wavelength) and $2 \times 0.6328$ $\mu$m, resp.

ductors, and magnetic excitations in antiferromagnetics. No attempt is made here to complete this list or to explain all the problems in detail [c]). The essential point is that useful information is obtained by an optical investigation, *i.e.* by measuring the reflection or transmission of a sample. With a conventional commercial double-beam instrument, reflectance or transmittance is obtained directly. In Fourier transform spectroscopy, where more radiative energy reaches the detector and a better signal-to-noise ratio is obtained, generally two measurements are carried out to obtain first the background spectrum $I_0(\tilde{\nu})$ (without sample) and

[c]) For further information, see Refs. [1–3]).

secondly the sample spectrum $I_s(\tilde{\nu})$ (with sample). The ratio of both spectra is then the reflectance $R$ or the transmittance $T$:

$$R(\tilde{\nu}) = \frac{I_s(\tilde{\nu})}{I_0(\tilde{\nu})} \quad \text{or} \quad T(\tilde{\nu}) = \frac{I_s(\tilde{\nu})}{I_0(\tilde{\nu})} \tag{4.14}$$

One example of a reflectivity measurement is shown in Fig. 26. The sample is $CdCr_2Se_4$ and the reflection spectra exhibit the typical reststrahlen bands caused by the lattice vibrations. Another example (Fig. 27) is a transmission measurement of an absorption line due to antiferromagnetic resonance. This also shows how, with spectroscopy in the extreme far-infrared, the signal-to-noise ratio is rather

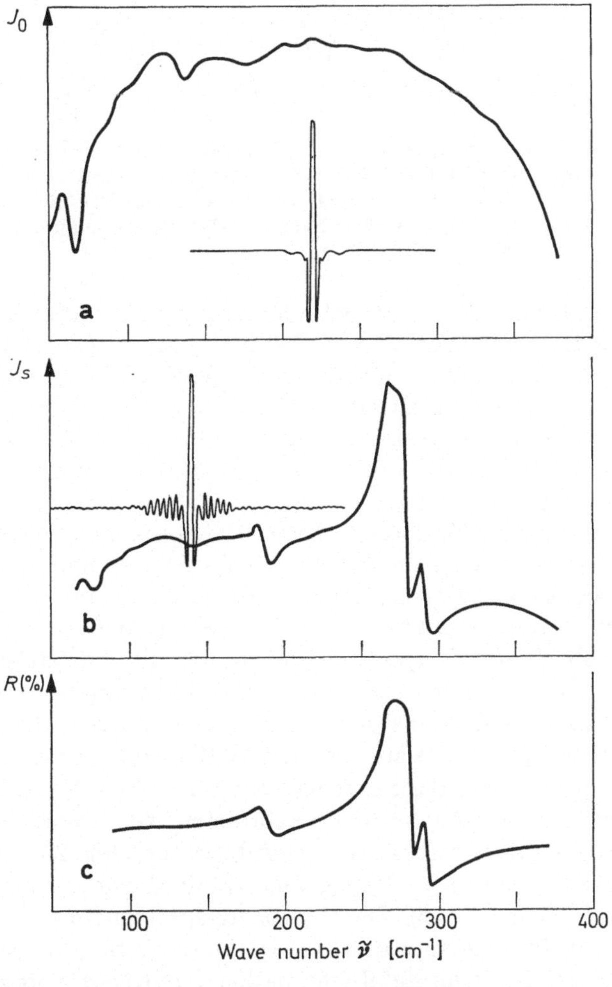


Fig. 26. Transmission and reflection study with a Fourier spectrometer. a) background spectrum and interferogram; b) sample spectrum and interferogram; c) ratio of both spectra (reflectivity of $CdCr_2Se_4$ in this example). — These data were obtained with a Beckman-RIIC Fourier spectrometer FS 720 with a Fourier transform computer FTC 300 attached to it

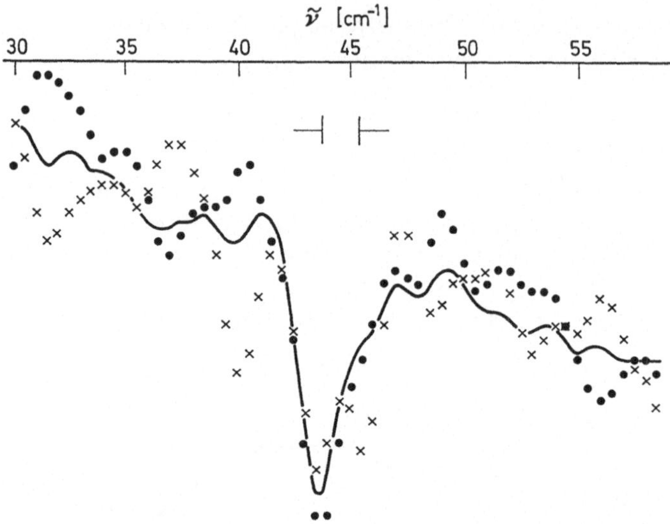


Fig. 27. Spectrum in the extreme far-infrared with an absorption line due to antiferromagnetic resonance (NiO, doped with 2% $Co^{2+}$) obtained by two single scans ( × and ·) and the average of several scans (——). These data were obtained with a Polytec FIR 30 Fourier spectrometer. The average was worked out by a Hewlett Packard HP 2100 A computer

limited. Here it is necessary to repeat a measurement several times and take the average of all the experimental data. With the rapid scan instruments in the middle-infrared, very often multiple scanning of the interferogram, *e.g.* 128 times, is used to improve the signal-to-noise ratio and the accuracy of the experimental data. The examples presented so far are taken from the field of solid state physics where the infrared spectroscopy plays an important role in determining the energies of various excitations. But it is also very well known that infrared spectroscopy is an valuable tool in chemical analysis, especially in the socalled "finger print" region. Here, modern rapid-scan Fourier spectrometers offer the possibility of an on-line analysis combining gas chromatography or liquid chromatography with infrared spectroscopy. And often an analysis does not mean to measure only an absorption spectrum and to indentify an certain substance via its absorption lines. In many cases, spectra have to be compared with each other. In the case of many biological systems for example, water is the only interesting solvent. But then, the spectra are obscured by the strong, broad infrared absorbance of water. Even under these extreme conditions, the infrared spectrum of the solved material can be determined by subtracting the absorbance spectrum of $H_2O$ as long as no total absorbance due to water occurs. Fig. 28 shows the aqueous solution infrared spectrum of hemoglobin obtained with a Fourier transform spectrometer (spectrum No. 2). The absorbance spectrum of the solvent is shown at the bottom of Fig. 28 (spectrum No. 1). An inspection of spectra No. 1 and No. 2 yields no detailed and useful information about hemoglobin. But when the difference of the two spectra is formed, the absorption lines characteristic for hemoglobin are clearly observed (spectrum No. 3 in Fig. 28: absorbance spectrum of hemoglobin in aqueous solution). It is evident that one of the absorption lines

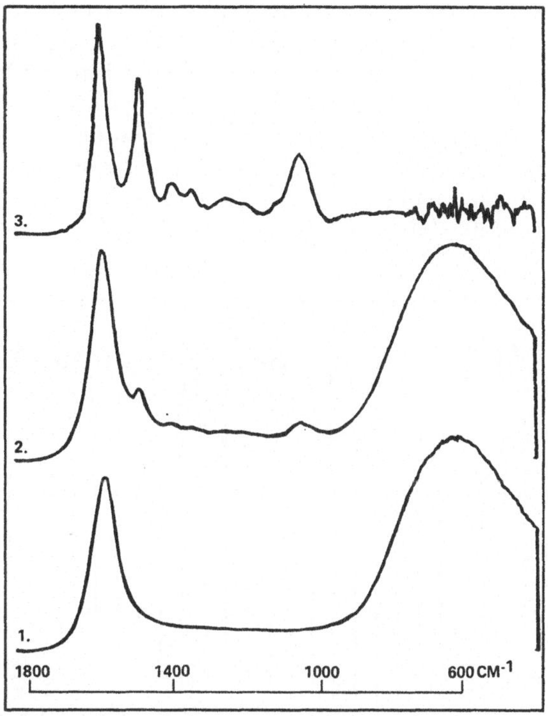


Fig. 28. Infrared spectrum of hemoglobin in aqueous solution — spectrum no 1: absorbance spectrum of $H_2O$, spectrum no. 2: absorbance spectrum of aqueous hemoglobin solution, and spectrum no. 3: absorbance spectrum of hemoglobin (difference of spectra no. 2 and no. 1). These spectra were obtained with a Digilab FTS 14 Fourierspectrometer (Data taken from Ref. [57c)])

due to the classical amide I and II modes (1657 and 1547 $cm^{-1}$) was hidden under the strong water absorption near 1600 $cm^{-1}$. Of course, such an analysis requires a rather high photometric accuracy. The possibility of multiple scanning has already been mentioned. And for the necessary electronic data processing, many manufacturers provide a software package with their instrument, in addition to the Fourier transform program. Among the additional programs, the following are useful for the analysis under consideration: double precision option, coadding and averaging several interferograms, conversion of transmittance into absorbance, and subtracting two (absorbance) spectra. In this context, it may be considered an advantage of Fourier spectroscopy that a computer is a necessary part of the spectrometer since a digital computer can be used for all calculations to which the spectra are subjected. And if the core of the computer is extended for this purpose by 8 K, for example, this is inexpensive in comparison to the cost of a complete Fourier spectrometer. At this point, it is perhaps worth mentioning that 8 K means 8000 words. In the computer language, 1 word with 12 or 16 bits means that a number between 0 and 4095 ($2^{12} - 1$) or 65535 ($2^{16} - 1$), respectively, can be stored at this place of the computer memory. For certain applications, double precision may be required. Double precision means that the word length is doubled to 24 or 32 bits, respectively [numbers from 0 to $(2^{24} - 1)$ or $(2^{32} - 1)$, resp.]. At the end of these considerations about chemical analysis and infrared spectroscopy, it remains to point out that the range 400–4000 $cm^{-1}$ is very important in this respect but that chemical investigations are not restricted to this range. For example, normal vibrations of macromolecules, molecules with heavy atoms, certain bending modes and torsional modes usually have frequencies below 500 $cm^{-1}$ in the far-infrared region [98].

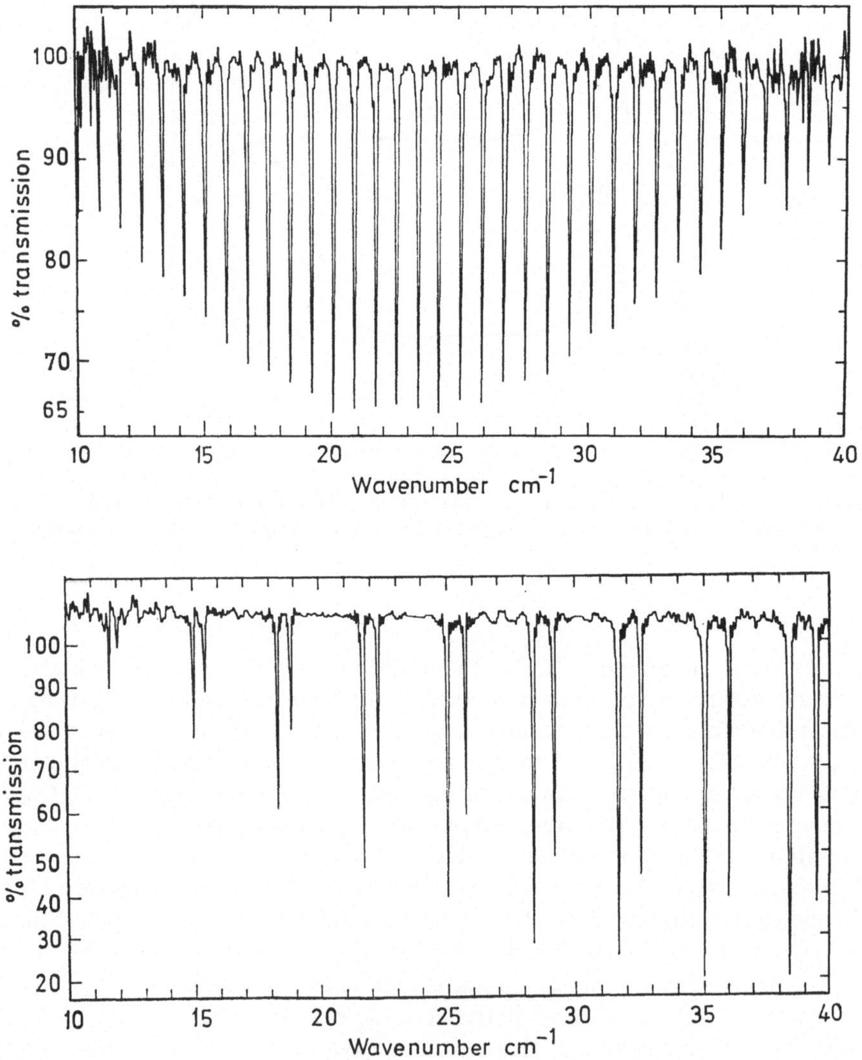


Fig. 29. Transmission spectrum of nitrous oxide ($N_2O$, upper half) and of nitric oxide (NO, lower half). Both spectra represent an average of three runs and were obtained at a pressure of 200 torr and a path length of 203 mm. The data were taken from Ref. [57a]. The instrument used was a somewhat modified Grubb Parsons Cube interferometer.

All the examples used here to illustrate Fourier transform spectroscopy in the infrared are more or less routine investigations, taken from the everyday work of spectroscopist. By contrast, the next samples will demonstrate some of the limits of capability of Fourier transform spectroscopy. J. W. Fleming and J. Chamberlain [57a)] have studied the transmission of nitrous oxide ($N_2O$) and nitric oxide (NO) in the extreme far infrared spectral range 10–40 $cm^{-1}$ (see Fig. 29). The absorption lines in the transmission spectra are due to rotational transitions. These spectra are rather impressive examples of extremely high resolution ($\Delta\tilde{\nu} \approx 0.05$ $cm^{-1}$, $R \approx 500$ at 25 $cm^{-1}$) which can be obtained only by means of Fourier spectroscopy in this range. It is not possible to obtain it by means of conventional grating spectroscopy with the same, or at least comparable, signal-to-noise ratio. It should be noted that in these experiments the resolution has been pushed that far that the limitations imposed on the resolution by the finite dimensions of the source and the finite aperture have to be taken into account (cf. Section 5.1). In many applications of spectroscopy to chemistry, very useful information about reactions may be gathered if the variation of spectral features in time can be observed. In this respect, R. E. Murphy *et al.*[57b)] have studied the spectral evolution of a $N_2/O_2$ gas mixture subjected to high-energy-electron irradiation. In this case, the electron gun was pulsed with a repetition frequency of about 80 Hz. For each pulse, the interferogram data were taken for a certain, fixed path difference. By repetition, the whole interferogram was scanned successively. The transformed emission spectra are presented in Fig. 30, which show clearly the change in time of the NO vibrational band at 1876 $cm^{-1}$ and of the vibration-rotation bands of $N_2O$ and $NO_2$ at 2200 and 1618 $cm^{-1}$, respectively. The time resolution is 50 $\mu$sec in these spectra. Our final example is taken from the pioneering work of P. and J. Connes, who applied Fourier transform spectroscopy to astrophysical investigations. They obtained spectra of the $CO_2$ rotation-vibration band ($\tilde{\nu} \approx 6500$ $cm^{-1}$ or $\lambda \approx 1.54$ $\mu$m) of Venus; these spectra are believed to demonstrate in a very excellent way the advantages of Fourier transform spectroscopy (Fig. 31). The data are compared to spectra of the $CO_2$ band obtained with a grating spectrometer in a laboratory and to one obtained with a conventional spectrometer from Venus.

## 4.7 Asymmetric Fourier Transform Spectroscopy

As everyone knows, the optical properties of a material are expressed in two optical constants, the refractive index $n$ and the absorption coefficient $\varkappa$. It is the purpose of spectroscopy to determine experimentally one or both of these optical constants as a function of frequency. This can be done by measuring reflection or transmission. If we were able to measure amplitudes or electrical fields (magnitude and phase) in an optical investigation, it would generally be possible to deduce both optical constants from one measurement of either reflection or transmission. However, we are only able to measure intensities where the magnitude of the field is determined and the phase information is lost. Thus, in general, from one item of information only one optical constant is obtained, and two measurements are necessary to determine both. There are a few exceptions to this rule, *e.g.* the

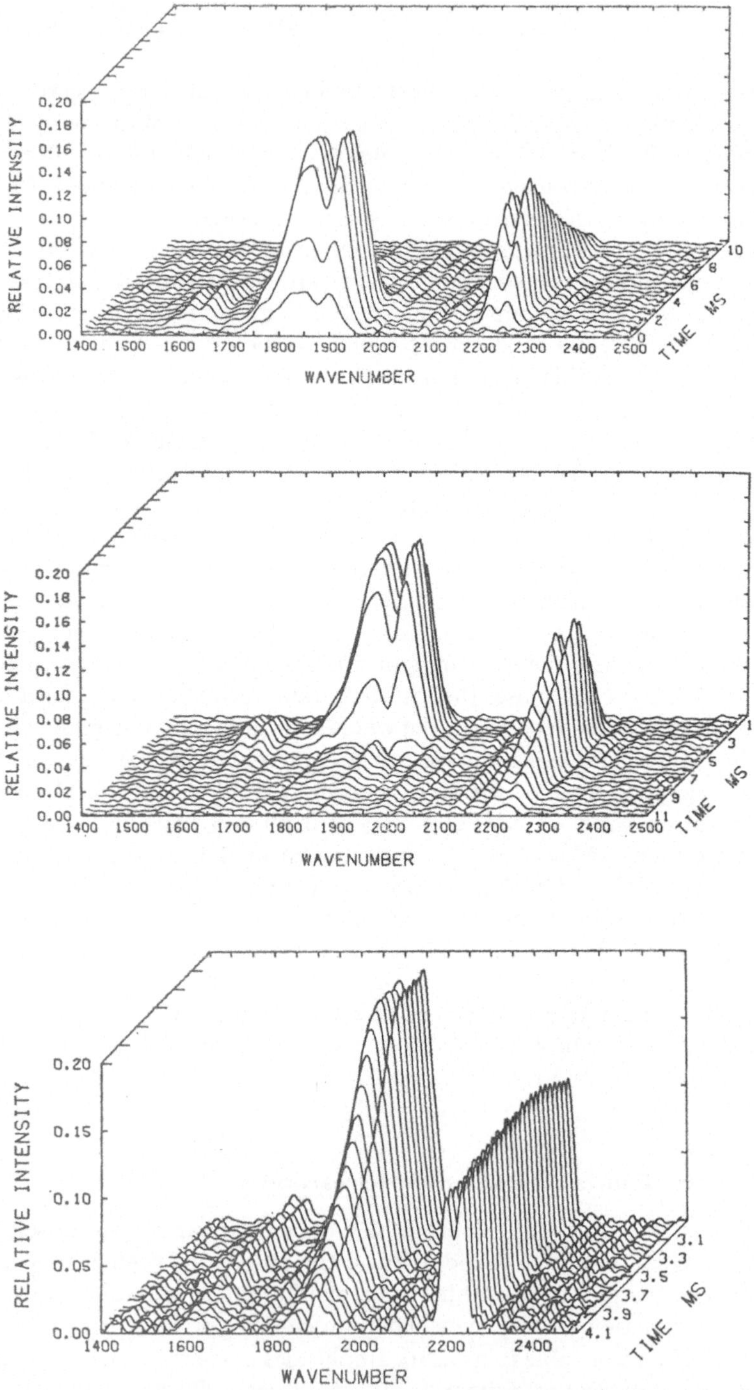


Fig. 30. Time resolved spectroscopy. The results are presented in a "three-dimensional" plot: intensity versus wave number (1400—2500 $cm^{-1}$) and time. a) time scale from front to rear (0—10 msec, $\Delta t = 500$ $\mu$sec), b) time scale from rear to front (11—1 msec, $\Delta t = 500$ $\mu$sec), c) a section of b) with enlarged time scale (4.1—3.1 msec, $\Delta t = 50$ $\mu$sec). The instrument used for this investigation was an Idealab IF-3 Fourier spectrometer with a PbSe (77 K) detector. The data were taken from Ref. [57b)]

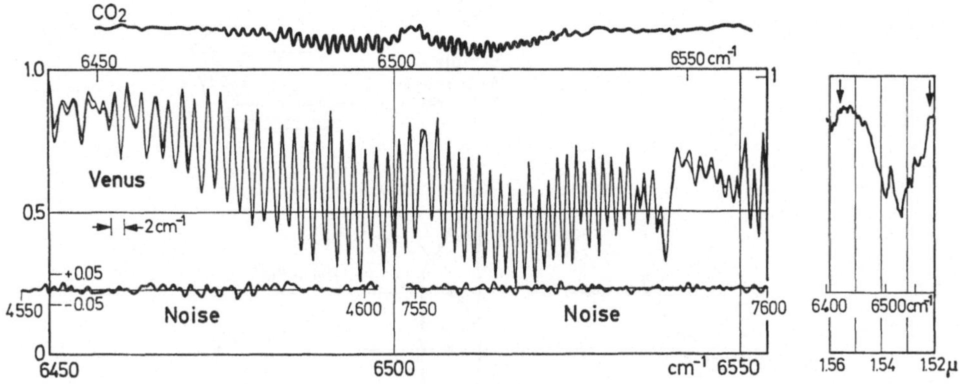


Fig. 31. Two superimposed spectra of the rotation-vibration band of $CO_2$ at about 6500 $cm^{-1}$ obtained from Venus by means of Fourier transform spectroscopy (left) and corresponding portion of the spectrum obtained by means of conventional spectroscopy (right). For comparison, a spectrum is shown which was obtained by conventional spectroscopy in a laboratory. Data taken from Ref. [58]

transmission spectrum from a plane-parallel plate where the absorption coefficient is determined by the intensity or energy loss in the sample (magnitude) and the refractive index by channeled spectra (phase). These statements are true for conventional spectroscopy as well as for Fourier spectroscopy when the sample is placed at the sample focus outside the arms of the Michelson interferometer. In contrast to conventional spectroscopy, however, Fourier transform spectroscopy offers the possibility of retaining the information about magnitude and phase and determining both optical constants from one measurement. This is achieved when the sample is placed in one arm of the Michelson interferometer. In this section, the principle is stated and some practical examples of this application of Fourier transform spectroscopy are given.

For transmission measurements where the sample is placed in one arm of the Michelson interferometer, a special optical arrangement is useful where the waves transmitted or reflected from the beam splitter to the mirrors and reflected by the mirrors travel at different heights. Fig. 32 is a schematic diagram of the arrangement developed by E. E. Bell [59 a,b], one of the pioneers in this field. The major advantage is that a sample is passed only once by the radiation. First, however, the sample is removed and a background interferogram is scanned and converted by the computer into the background spectrum, $I_0(\tilde{\nu})$.

When the sample is put into one of the partial beams, the magnitude of the wave in this beam is reduced according to the transmission coefficient of the sample. In addition, a phase shift is introduced by the change in optical path length due to the sample. For monochromatic radiation, we have to write for the wave in the arm of the fixed mirror, instead of Eq. (2.3a):

$$E_1 = \sqrt{T(\tilde{\nu})}\, E_0\, e^{i(2\pi\nu t - 2\pi\tilde{\nu} r_1 - \varphi(\tilde{\nu}))} \,. \tag{4.15}$$

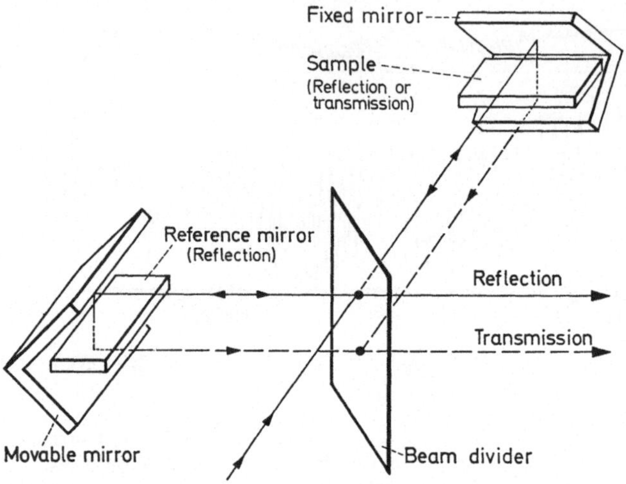


Fig. 32. Optical arrangement (principle) for asymmetric Fourier transform spectroscopy. For further details see Ref. [60]

Here $T(\tilde{\nu})$ is the power or intensity transmittance of the sample and $\varphi(\tilde{\nu})$ is the phase shift introduced by it. The wave reflected at the movable mirror can still be written in the form of Eq. (2.3b). The interferogram with the sample in one arm of the Michelson interferometer is then for monochromatic radiation

$$I_s(s) = \frac{1}{2}\sqrt{\frac{\varepsilon_0}{\mu_0}}\,|E_1 + E_2|^2 \tag{4.16}$$

$$= I_0(\tilde{\nu})[1 + T(\tilde{\nu}) + 2\sqrt{T(\tilde{\nu})}\cos(2\pi\tilde{\nu}s - \varphi(\tilde{\nu}))]\,,$$

where $I_0(\tilde{\nu}) = \frac{1}{2}\sqrt{\frac{\varepsilon_0}{\mu_0}}\,E^2$ and $s = r_2 - r_1$. For a continuous spectrum, we have to add up all the contributions from spectral elements of wave number $\tilde{\nu}$ and obtain for the oscillatory part of the sample interferogram

$$\tilde{I}_s(s) = 2\int_0^\infty \sqrt{T(\tilde{\nu})}\,I_0(\tilde{\nu})\cos(2\pi\tilde{\nu}s - \varphi(\tilde{\nu}))\,d\tilde{\nu}\,. \tag{4.17}$$

The grand maximum of the sample interferogram is shifted to higher wave numbers than the background interferogram (Fig. 33). If the reflectivity $R$ of the sample is small and if the absorption coefficient $\varkappa$ is small compared to the refractive index $n$, the magnitude and the phase of the transmission coefficient may be approximated by

$$T(\tilde{\nu}) = (1 - R)^2 e^{-4\pi\varkappa d\tilde{\nu}} \qquad \varphi(\tilde{\nu}) = 2\pi(n - 1)\,d\tilde{\nu}$$

$$\text{with } R \approx \left(\frac{1-n}{1+n}\right)^2 . \tag{4.18}$$

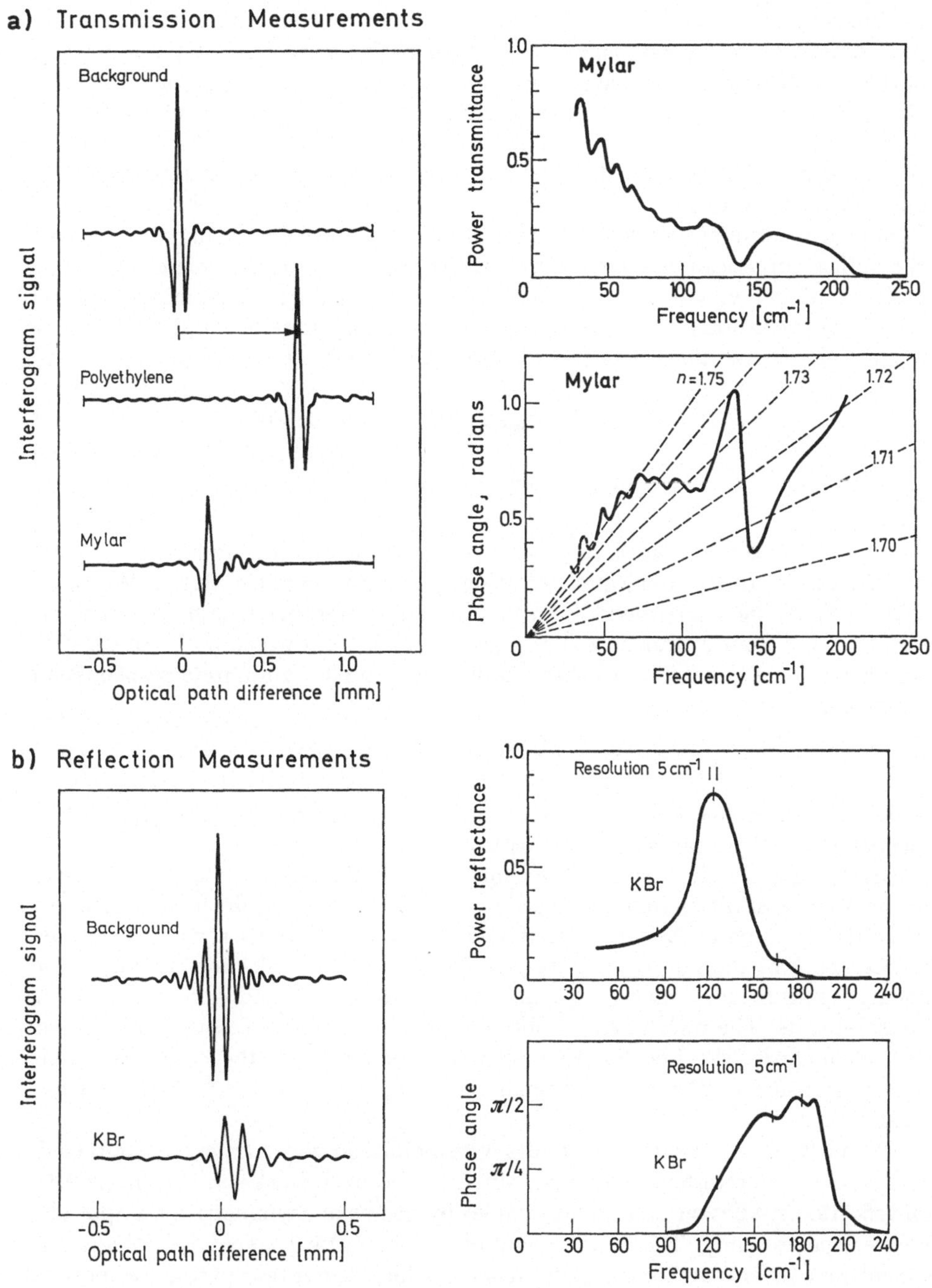


Fig. 33. Transmission (a) and reflection measurements (b) by means of asymmetric Fourier transform spectroscopy. The samples are polyethylene (PET) and mylar (transmission), and KBr (reflection). The different curves show background and sample interferograms as well as power transmittance (reflectance) and phase angle spectra. Data taken from Ref. [59]

From these relations, it is clear that the sample interferogram is shifted by the path difference corresponding to the optical path $nd\tilde{\nu}$ in the sample minus the path in air $d\tilde{\nu}$ replaced by the sample. For the case of nearly constant $n$ (see Fig. 33, PET sample), the interferogram is simply shifted by this amount and is still symmetric. In general, however, $n$ and $\varkappa$ depend on the wave number $\tilde{\nu}$. The phase shifts are then different for different wave numbers, and the resulting interferogram is asymmetrically distorted (see Fig. 33, mylar sample). In this kind of Fourier transform spectroscopy the interferogram is usually asymmetric because the optical properties of the two arms of the Michelson interferometer are asymmetric and not equal. This special application is called "asymmetric Fourier transform spectroscopy". The properties of the sample are deduced by means of a cosine and a sine Fourier transform from the sample interferogram

$$\begin{aligned} \int \bar{I}_s(s) \cos(2\pi\tilde{\nu}s)\, ds &\approx \sqrt{T(\tilde{\nu})}\, I_0(\tilde{\nu}) \cos\varphi(\tilde{\nu}) \\ \int \bar{I}_s(s) \sin(2\pi\tilde{\nu}s)\, ds &\approx \sqrt{T(\tilde{\nu})}\, I_0(\tilde{\nu}) \sin\varphi(\tilde{\nu})\,. \end{aligned} \tag{4.19}$$

Eq. (4.19) simply shows the basic relationship and the influence of finite interferogram, apodization, and digitizing is not considered in detail. We recall that $I_0(\tilde{\nu})$ is the background intensity already determined, and the essential results of the Fourier transform are $T(\tilde{\nu})$ and $\varphi(\tilde{\nu})$, from which both optical constants can be evaluated. In other words, the complex amplitude transmission coefficient

$$\sqrt{T(\tilde{\nu})}\quad e^{-i\varphi(\tilde{\nu})}$$

is determined in this kind of Fourier spectroscopy. Therefore, it is often called "amplitude Fourier spectroscopy".

In the case of reflection measurements, the sample replaces one of the mirrors in the Michelson interferometer (see Fig. 32). The reference mirror is assumed to be 100% reflecting in the far-infrared, and in the sample interferogram the power reflectance $R$ of the sample and the phase shift $\psi$ at the reflection (usually $\pi$ for nonabsorbing media with $n > 1$) take over the role of $T$ and $\varphi$ in transmission measurements. The interferogram obtained in this case is also somewhat shifted and asymmetric (see Fig. 33, KBr sample). By means of the cosine and sine Fourier transforms, $R$ and $\psi$, and finally $n$ and $\varkappa$, are evaluated from the experimental data.

Recently, first experimental results were published which have been obtained with a Michelson interferometer especially designed by J. Gast and L. Genzel[60a, b] for reflection studies on small solid samples by means of asymmetric or amplitude Fourier spectroscopy. The main advantage of the optical layout is that sample and reference mirror are located at focal points which do not take part in the motion to produce the path difference in the interferometer. Therefore, these foci can be placed inside a cryostat that allows the sample to be cooled. Another recent development in this field is concerned with the difficulty that, when studying the reflectivity of solids, the determination of the phase depends strongly on the exact positioning of the sample mirror instead of the background mirror. This is overcome when the sample is partly aluminised and used by itself as a background mirror. Sample and background interferogram are obtained with the same setup by dividing the field of view [61].

But the amplitude or asymmetric Fourier spectroscopy has not been applied only to solids but also to liquid samples and gases. Here too, it has been proved an important help to obtain the two optical constants n and $\varkappa$ simultaneously [62a,b]. In gases, for example, the anomalous dispersion of the refractive index in the neighbourhood of rotational absorption lines was determined experimentally by this method.

## 5. Advantages and Disadvantages

In this section we comment on the advantages and disadvantages of Fourier transform spectroscopy as compared with conventional spectrocsopy. The most important of the principal advantages of Fourier spectroscopy are the Jacquinot and the Fellgett advantages, often cited in the literature, and called after the pioneers in this field who first pointed them out. Even for practical spectroscopy, it is good to know what these advantages really mean. Retrospectively, we also discuss whether the Fourier transform, the use of a computer etc., is a great disadvantage of the method or a tolerable one. Next, it is important to have some idea of the requirements in Fourier transform spectroscopy with respect to the necessary mechanical precision and alignment of the optical system, also filtering and sensitivity. Then some hints will be given about possible errors and their suppression. In many cases, if we know about a certain error, it can be suppressed by introducing a special factor in the mathematical treatment, *i.e.* the Fourier transform. Finally, we discuss the average noise level in the spectrum with a given signal-to-noise ratio in the interferogram. For the spectroscopist, the answer to this question is helpful in determining the optimum length of an interferogram, *i.e.* the optimum resolution for a measurement.

### 5.1 Principal Advantages and Disadvantages

In the discussion of the basic principles of Fourier transform spectroscopy a number of advantages and disadvantages have already been mentioned. For example, a much greater portion of the radiant power emitted by the source reaches the detector in a Michelson interferometer than is the case in a grating spectrometer. In a naive and qualitative way, this is due to the separation of the radiation into spectral elements in the grating spectrometer, but a quantitative comparison shows that, in addition to this, there is another principal advantage of the Michelson interferometer, the so-called Jacquinot advantage. On the other hand, a better signal-to-noise ratio is required in Fourier transform spectroscopy before the finer details of the spectrum can be extracted from the interferogram.

Let us consider now all these aspects in more detail and in a quantitative way. Generally, the radiant power utilized from the source in a Michelson interfero-

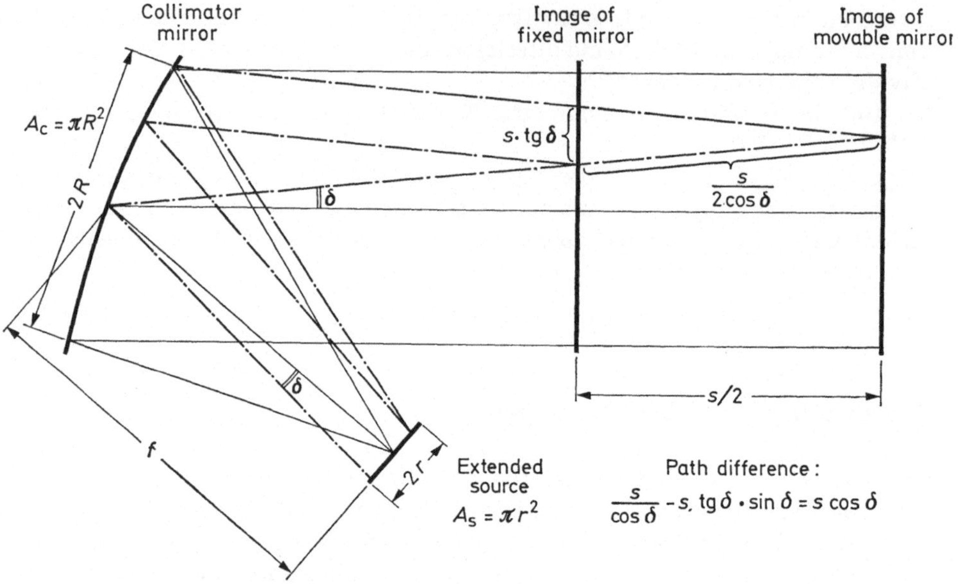


Fig. 34. Extended source problem in a Michelson interferometer

meter is proportional to the brightness $B$ and the area $A_s$ of the source and the solid angle $\Omega_c$ subtended by the collimator mirror (Fig. 34) [32]:

$$P = B \cdot A_s \cdot \Omega_c = B \frac{A_s \cdot A_c}{f^2} . \tag{5.1}$$

The solid angle may be expressed by the area $A_c$ of the collimator mirror and its focal length $f$. Thus we arrive at the more symmetric form of Eq. (5.1). In the literature, the quotient $A_s/f^2$ is often expressed as the solid angle $\Omega_s$ which the source subtends from the collimator mirror. In a lossfree and ideal optical system, the product $A \cdot \Omega$ is constant throughout the system according to the rules of geometrical optics [32]. Therefore, it is a suitable quantity to describe the flux throughput of an optical system. Originally, P. Jacquinot focused attention on this fact. He called this quantity "étendu" or throughput [20, 63]. It is

$$E_M = A_c \cdot \Omega_s = \frac{A_c \cdot A_s}{f^2} \tag{5.2}$$

for the Michelson interferometer. If we want to evaluate the "étendu" for a grating spectrometer, we have to replace the area $A_s$ of the source by the effective source area, *i.e.* the area of the slits $w \cdot h$, where $w$ and $h$ are the width and the height of the slit. Then we obtain

$$E_G = A_c \cdot \frac{w \cdot h}{f^2} . \tag{5.3}$$

Next, let us consider that the finite dimensions of the source impose a limit to the resolving power of the Michelson interferometer in the same way as the finite slit width does for the grating spectrometer. The finite size of the source means that also rays enter the interferometer which are inclined by an angle $\delta$ to the optical axis (see Fig. 34); hence, a displacement $s/2$ of the movable mirror produces the path difference $s \cdot \cos\delta$ instead of $s$. The total interferogram is the sum of the contributions from all points of the (circular!) source [2,4]:

$$I(s) = 4\pi \int_0^\infty \int_0^{\delta_{\max}} I(\tilde{\nu})[1 + \cos(2\pi\tilde{\nu}s\cos\delta)] \sin\delta \, d\delta \, d\tilde{\nu}$$

$$= 2\pi\frac{r^2}{f^2} \int_0^\infty I(\tilde{\nu}) \left[ 1 + \frac{\sin\left(\pi\tilde{\nu}s\frac{r^2}{2f^2}\right)}{\left(\pi\tilde{\nu}s\frac{r^2}{2f^2}\right)} \cos\left(2\pi\tilde{\nu}s\left\langle 1 - \frac{r^2}{4f^2}\right\rangle\right)\right] d\tilde{\nu} \quad (5.4)$$

$$\text{with } \delta_{\max} = \arccos\left(1 - \frac{r^2}{2f^2}\right)$$

where $r$ is the radius of the source and $f$ is again the focal length of the collimator. The Fourier transform of this interferogram, neglecting here all effects of finite interferogram, apodization and digitizing, gives

$$I_{\text{obs}}\left(\tilde{\nu}\left[1 - \frac{r^2}{4f^2}\right]\right) = \int_{\tilde{\nu}\left(1-\frac{r^2}{4f^2}\right)}^{\tilde{\nu}\left(1+\frac{r^2}{4f^2}\right)} I(\tilde{\nu}')\, d\tilde{\nu}' . \quad (5.5)$$

Eq. (5.5) tells us that the effect of the finite size of the source is twofold:

1) The "true" wave number $\tilde{\nu}$ is observed at a lower wave number $\tilde{\nu}\left(1 - \frac{r^2}{4f^2}\right)$.
2) At this wave number, we find contributions from the range $\tilde{\nu}\left(1 - \frac{r^2}{4f^2}\right) \leqq \tilde{\nu}' \leqq \tilde{\nu}\left(1 + \frac{r^2}{4f^2}\right)$. This means a finite resolution $\Delta\tilde{\nu} = \tilde{\nu}\frac{r^2}{2f^2}$.

Thus, the resolving power from the finite area of the source in the Michelson interferometer is

$$R_{\text{M}} = \frac{\tilde{\nu}}{\Delta\tilde{\nu}} = 2\frac{f^2}{r^2} . \quad (5.6)$$

For a diffraction grating, on the other hand, the basic formula [Eq. (2.1)] with $\varphi_{\text{d}} = \varepsilon + \theta$ and $\varphi_{\text{i}} = \varepsilon - \theta$ (Fig. 35) reads

$$n \cdot \lambda = 2g \cos\varepsilon \sin\theta \quad (5.7)$$

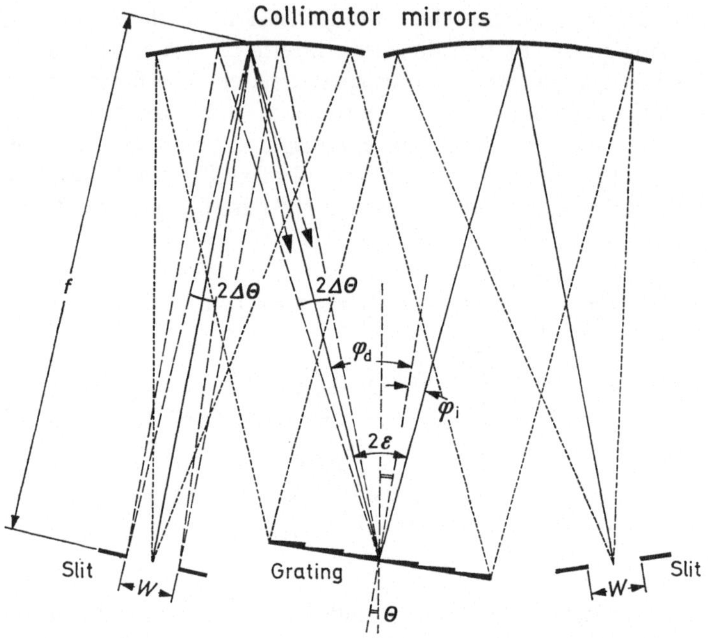


Fig. 35. Limitation of the resolution of a grating spectrometer by the finite width of the slits

From Eq. (5.7) we obtain the resolution of a grating spectrometer as limited by the slit width (see also Section 3.1) [3])

$$\Delta \tilde{\nu} = -\frac{d\tilde{\nu}}{d\theta}\,\Delta\theta = \tilde{\nu}\,\mathrm{ctg}\,\theta\,\frac{w}{2f} \tag{5.8}$$

where the slit width $w$ is equal to $2f \cdot \Delta\theta$ from simple geometric considerations. Thus, the energy-limited resolving power of the grating spectrometer becomes

$$R_G = \frac{\tilde{\nu}}{\Delta\tilde{\nu}} = \frac{2f}{w}\,\mathrm{tg}\theta\,. \tag{5.9}$$

The throughput $E$ and the resolving power $R$ are quantities characteristic for a given instrument. When comparing scanning times and signal-to-noise ratios, however, it is advisable to include the brightness of the source and to consider the actual power flux through the instrument see [Eq. (5.1)]. For simplicity, let us assume that the brightness of the source may be approximated by that of a black-body source at a certain temperature $T$. Then, for the far-infrared spectral range, the relation $hc\tilde{\nu} \ll kT$ is assumed to hold and we can use the Rayleigh-Jeans law [32])

$$I(\tilde{\nu})\,d\tilde{\nu} = 2ckT\tilde{\nu}^2 d\tilde{\nu}\,. \tag{5.10}$$

For the Michelson interferometer, the radiation is a broad band from 0 to a maximum wave number $\tilde{\nu}_{max}$:

$$B_M = \int_0^{\tilde{\nu}_{max}} I(\tilde{\nu})\, d\tilde{\nu} = \frac{2}{3}\, ckT\tilde{\nu}_{max}^3$$

$$\text{and} \quad P_M = B_M \cdot E_M = \frac{2}{3}\, ckT\,\tilde{\nu}_{max}^3\, \frac{A_c \cdot A_s}{f^2}. \tag{5.11}$$

For the grating spectrometer, on the other hand, the radiation is limited to the bandwidth $\Delta\tilde{\nu}$ as given by the resolution of the instrument:

$$B_G = I(\tilde{\nu}) \cdot \Delta\tilde{\nu} = 2\, ckT\tilde{\nu}^2 \Delta\tilde{\nu}$$

$$\text{and with} \qquad \Delta\tilde{\nu} = \tilde{\nu}\, \mathrm{ctg}\, \theta\, \frac{w}{2f} \tag{5.12}$$

$$P_G = B_G \cdot E_G = 2\, ckT\tilde{\nu}^3 A_c\, \frac{w^2 h}{2f^3}\, \mathrm{ctg}\, \theta\,.$$

From Eqs. (5.11) and (5.12), we see that $B$ is larger for the Michelson than for the grating instrument. It will be shown below that, as the throughput $E$ (Jacquinot advantage) is larger for the Michelson, $P$ is much larger.

In order to demonstrate the significance and meaning of these formulae, let us insert some typical figures for far-infrared instruments and compare the results. For a Michelson interferometer, the following values are taken as typical:

| | |
|---|---|
| Collimator area | $A_c = 176.7\ \mathrm{cm}^2\ (R = 7.5\ \mathrm{cm})$ |
| Focal length | $f = 30\ \mathrm{cm}$ |
| Source (radius) | $r = 0.5\ \mathrm{cm}$ |
| Maximum feasible path difference | $s_{max} = 25\ \mathrm{cm}$ |
| Maximum wave number | $\tilde{\nu}_{max} = 130\ \mathrm{cm}^{-1}$ |

and for a grating spectrometer:

| | |
|---|---|
| Collimator area | $A_c = 706.9\ \mathrm{cm}^2\ (R = 15\ \mathrm{cm})$ |
| Focal length | $f = 92\ \mathrm{cm}$ |
| Grating constant | $g = 0.0300\ \mathrm{cm}$ |
| Number of grooves | $N = 1000$ |
| Slit width | $w = 0.20\ \mathrm{cm}$ |
| Slit height | $h = 1.50\ \mathrm{cm}$ |
| For $\tilde{\nu} = 50\ \mathrm{cm}^{-1}$ in 1st order | $\theta = 19.9°$ |

For both instruments $T = 3000$ K has been assumed to be the average temperature of the source [71]. From the values listed above, the interesting quantities $E$, $R$ and $P$ were computed. The results for the Michelson interferometer and the grating spectrometer are listed below for comparison:

| | Michelson interferometer | Grating spectrometer |
|---|---|---|
| Throughput | $E_M = 0.154$ cm$^2$·ster | $E_G = 0.025$ cm$^2$·ster |
| Power flux | $P_M = 2.80 \cdot 10^{-4}$ W | $P_G = 2.33 \cdot 10^{-8}$ W |
| Resolving power from finite dimensions of source or slits | $R_M = 7200$<br>$\Delta\tilde{\nu} = 0.007$ cm$^{-1}$ [1]) | $R_G = 333.3$<br>$\Delta\tilde{\nu} = 0.15$ cm$^{-1}$ [1]) |
| Resolving power from number of grooves or maximum path difference | $R_M = 1250$ [1])<br>$\Delta\tilde{\nu} = 0.04$ cm$^{-1}$ | $R_G = 1000$<br>$\tilde{\nu}\Delta = 0.05$ cm$^{-1}$ [1]) |

[1]) at $\tilde{\nu} = 50$ cm$^{-1}$.

The resolving power from interference effects (maximum path difference or number of grooves) is nearly equal, in both cases for the typical figures chosen here, but the resolving power derived from the finite dimensions of the source or of the slits is very different for the two instruments. The value of the Michelson interferometer is about 20 times larger than that of the grating spectrometer. These results substantiate the statements in Sections 3.1 and 3.2 that resolution in Fourier transform spectroscopy is limited by interference effects [$R = \tilde{\nu} s_{max}$, see Eq. (2.27)], while that of the grating spectrometer is energy-limited [slit width, Eq. (5.9)]. However, there have been examples of high resolution Fourrier spectroscopy where the finite aperture and the finite dimensions of the source have a perceptable influence on the results. The transmission spectra of $N_2O$ and NO which were obtained by J. W. Fleming and J. Chamberlain [57a] are already presented in Section 4.6 (see Fig. 29). Their nominal resolution has been $\Delta\tilde{\nu} = 0.05$ cm$^{-1}$ as calculated from the maximum path difference [cf. Eq. (2.20)]. The solid angle $\Omega_s$ which the source subtends from the collimator mirror was in their experiment [cf. Eq. (5.2)]

$$\Omega_s = \frac{\pi r^2}{f^2} \approx 0.0035 \approx 3 \cdot 10^{-4} \cdot 4\pi \text{ sterad.}$$

According to Eq. (5.5), the resolution due to the finite size of the source would be

$$\Delta\tilde{\nu} = \tilde{\nu}\frac{r^2}{2f^2} = \tilde{\nu}\frac{\Omega_s}{2\pi} \approx \tilde{\nu}\,(0.0006) \tag{5.5a}$$

in this case. At 40 cm$^{-1}$, its value is $\Delta\tilde{\nu} = 0.025$ cm$^{-1}$ which is still somewhat smaller than the resolution due to the maximum path difference. But it is of the same order of magnitude. Moreover, the authors were able to determine the peak positions of the absorption lines with an accuracy of about 0.003 $cm^{-1}$. And their experimental values of the frequencies of the rotational transitions in $N_2O$ and NO differed systematically from the true values by a factor

$$\tilde{\nu}_{\text{true}} = \tilde{\nu}_{\text{obs}} \cdot (1.00035)$$

[see Eq. (5) in Ref. [57a)]]. This wave number shift is to be regarded the shift resulting from the finite size of the source [see Eq. (5.5) and its discussion]

$$\tilde{\nu}_{\text{obs}} = \tilde{\nu}_{\text{true}} \left(1 - \frac{r^2}{4f^2}\right) = \tilde{\nu}_{\text{true}} \left(1 - \frac{\Omega_s}{4\pi}\right). \tag{5.5b}$$

Inserting the value of $\Omega_s$, we obtain from this relation ($\Omega_s/4\pi \ll 1$)

$$\tilde{\nu}_{\text{true}} = \tilde{\nu}_{\text{obs}} \left(1 + \frac{\Omega_s}{4\pi}\right) \approx \tilde{\nu}_{\text{obs}}\,(1.0003)$$

This estimate of the wave number shift agrees reasonably with the experimental value.

The throughput of the Michelson interferometer is about 6 times larger than that of the grating spectrometer. It would be larger by another factor of 20 if one could assume equal energy-limited resolution for both instruments. The values of the throughput, quoted here are typical for spectrometers designed and built in a laboratory for the very far-infrared. A grating instrument with collimator mirrors of 30 cm diameter and with a focal length of 92 cm is a rather huge instrument. The values of the throughput of commercially available interferometers range from 0.03 to 0.2 $cm^2 \cdot$ ster (cf. Section 6.3). Especially instruments for the near- and middle-infrared have a somewhat lower throughput in order to minimize aberrations and to obtain a small and handy instrument. And a grating spectrometer with the dimensions of such a Michelson interferometer would have a throughput much smaller than the value quoted in the comparison.

It should be noted further that an increase in resolution is easily achieved in this case by increasing the maximum path difference and the scanning time. The power flux is not influenced by an increase of $s_{max}$. However, there will be an increase in noise, as we shall see later. An increase in resolution means for a grating instrument a reduction of slit width and hence, a reduction of the power flux, which is proportional to the square of the slit width [see Eq. (5.12)]. It also seems worth mentioning that the Jacquinot or throughput advantage exists not only in the Michelson interferometer but also in other instruments, e.g. a Fabry-Perot interferometer.

The other principal advantage which applies to Fourier transform spectroscopy is the "multiplex" or "Fellgett" advantage [21,64)]. It was P. Fellgett who first pointed out that there is an advantage when the data in all elements of a spectrum are obtained simultaneously instead of being measured for each element successively. In Fourier transform spectroscopy, the radiation in the Michelson interferometer is not separated into spectral elements. The interferogram contains the information about all these elements and, in the Fourier transform with a digital computer, this information is extracted from the experimental data for all spectral elements simultaneously. The multiplex advantage does not always apply to Fourier transform spectroscopy. If the Fourier transform is performed in the analogoue way according to the principle of interference modulation (see Section 4.2) and if only one phase-sensitive rectifier and low-pass filter are used (see Fig. 19), we obtain only the information of a single spectral element and we cannot profit from the multiplex advantage. This is also true for a conventional grating instrument when the radiation of one spectral element is transmitted to the detector, but when a photographic plate is used in the plane of the exit slit, the multiplex advantage is also effective in this case of conventional spectroscopy.

In order to derive a more quantitative formulation of the Fellgett advantage, let us consider the analogue Fourier transform spectrometer, as described in Section 4.2. There, it is possible to compare experiments with and without multiplexing for the same instrument. For different instruments, the comparison may be obscured by other advantages or disadvantages, *e.g.* the Jacquinot advantage. Let us assume that we have to measure in a certain time $T$ the intensity of $M$ spectral elements at wave number $\tilde{\nu} = m \Delta\tilde{\nu}$ of width $\Delta\tilde{\nu}$ in the range $0 \leqq \tilde{\nu} \leqq M\Delta\tilde{\nu}$ or $0 \leqq m \leqq M$. If the instrument is equipped with $M$ phase-sensitive rectifiers (see Fig. 19), we can measure the intensity for all wave numbers $m\Delta\tilde{\nu}$ simultaneously, the measuring time for each spectral element being the total time $T$. The $M$ phase-sensitive rectifiers are tuned to the modulation frequencies $2v_0 m\Delta\tilde{\nu}$ [see Eq. (4.3)] corresponding to the wave number $m\Delta\tilde{\nu}$. If the instrument is equipped with only one phase-sensitive rectifier, the intensity of only one spectral element can be measured at a time: successively, the one phase-sensitive rectifier is tuned to the different modulation frequencies $2v_0 m\Delta\tilde{\nu}$, and the intensity $I(m\Delta\tilde{\nu})$ is obtained. In this case, we require the measuring time $T/M$ for each single spectral element.

Now, the quantitative aspect of the multiplex advantage is that a better signal-to-noise ratio is achieved for the long time $T$ in comparison to the short time $T/M$. In far-infrared spectroscopy, the noise is mostly detector noise, independent of the signal. Under optimum conditions, the time constant $\tau = RC$ of the low-pass filters will be chosen proportional to $T$ ur $T/M$, in partice $\tau \sim \frac{1}{3} T$ or $\tau \sim \frac{1}{3} \frac{T}{M}$. (cf. no. 5 in Section 4.6). Now, the root mean square value $\sqrt{\overline{N^2}}$ (RMS value) of the noise voltage is proportional to $\sqrt{\Delta f} = \sqrt{\frac{1}{\tau}}$ from both technical and statistical arguments. For the multiplex case, we have

$$(\sqrt{\overline{N^2}})_{\text{Mult}} \sim \sqrt{\frac{1}{T}} \quad \text{and} \quad \left(\frac{S}{\sqrt{\overline{N^2}}}\right)_{\text{Mult}} \sim \sqrt{T}\,, \tag{5.13}$$

where $S$ is the signal and $(S/\sqrt{\overline{N^2}})$ the signal-to-noise ratio. For the non-multiplex case, we obtain

$$(\sqrt{\overline{N^2}})_{\text{NM}} \sim \sqrt{\frac{M}{T}} \quad \text{and} \quad \left(\frac{S}{\sqrt{\overline{N^2}}}\right)_{\text{NM}} \sim \sqrt{\frac{T}{M}}. \tag{5.14}$$

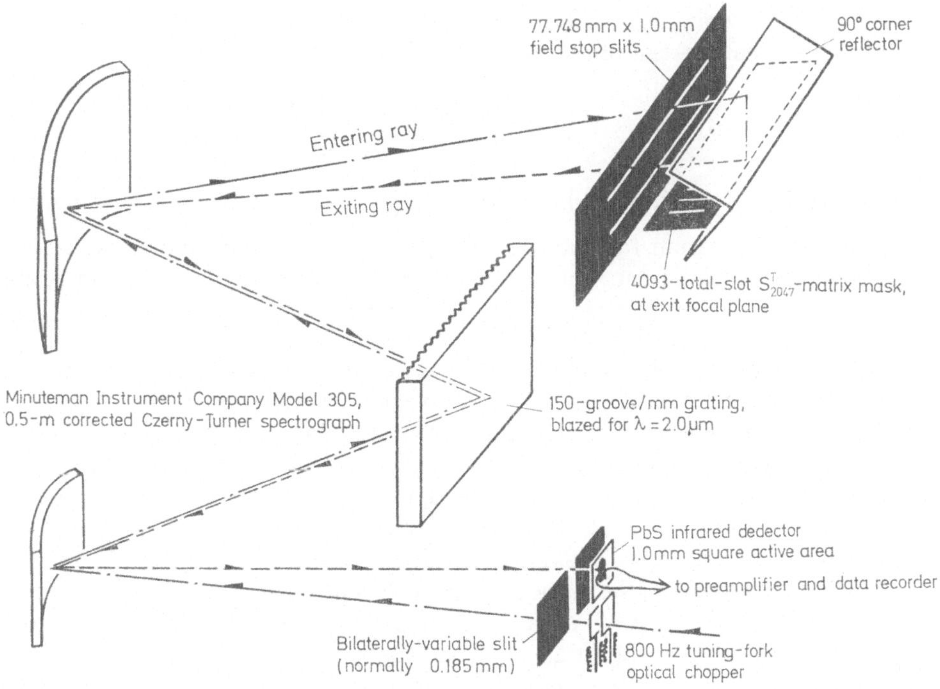


Fig. 36. Optical layout of a 2047-slot Hadamard transform spectrometer (taken from Ref.[77])

The results mean that a factor of $\sqrt{M}$ is gained in the signal-to-noise ratio for the multiplex case. In this context, it should be noted that a long wave number range can be scanned with a Fourier spectrometer in quite a short time, especially with a rapid scan Fourier spectrometer. The multiplex advantage is considered so important that attempts have been made to realize it also for conventional grating spectroscopy in the infrared region where the photographic plate cannot be used as a detector. A method was developed which is known as Hadamard transform spectroscopy[72–77]. Its principle can be described as follows: In a grating spectrometer, the intensity of all the spectral elements is distributed in the focal plane of the exit slit. Instead of a single slit, a multislit mask consisting of transparent and opaque slots is used through which the dispersed light is passed to obtain an encoded signal with all the information from a broad spectral range (cf. Fig. 36). The signal at the detector is the sum of all the light transmitted by the transparent slots. Now it is obvious that a decoding for $N$ spectral elements is only possible if $N$ independent measurements have been performed. Please note, that in Fourier spectroscopy these $N$ independent measurements are the interferogram values $I(n\Delta s)$ at $N$ different values of path difference. In Hadamard spectroscopy the independent informations are obtained by using a mask with $2N-1$ slots but illuminating only $N$ of them at a time. Then the mask can be shifted $N$ times by one slot and each time a different portion of $N$ slots is illuminated (see Fig. 37).

In this case, the code is a cyclic $N \cdot N$ matrix the elements of which are "1" (transparent) or "0" (opaque). In Fourier transform spectroscopy, the corresponding encoding system are the phase factors

$$\cos(2\pi\tilde{\nu}s) = \cos\left(\pi\frac{m \cdot n}{N}\right)$$

for $\tilde{\nu} = m\Delta\tilde{\nu}$, $s = n\Delta s$, and $\Delta\nu \cdot \Delta s = \frac{1}{2N}$ (see Sections 2.2 and 4.3). To extract the spectral data from the interferogram, the encoded information, we have to use the Fourier transform. This transform is the inverse of the encoding system. It consists of the same phase factors $\cos\left(\pi\frac{mn}{N}\right)$ [see Eq. (4.8) and Figs. 21 and 22 in Sections 4.3 and 4.4]. In Hadamard spectroscopy, too, a mathematical treatment of the recorded data is necessary for decoding and for extracting the spectral data. Here, this procedure is the application of a matrix inverse to the encoding matrix in Fig. 37. The latter usually is chosen in such a way, that on the average, one half of the slots is transparent and the other half opaque. From these considerations, it is evident, that Hadamard transform spectroscopy can profit from the multiplex or "Fellgett" advantage. But also the throughput can be enlarged in this case if a second slot mask is used in the entrance focal plane instead of a single slit [78] (Note, that this case has not been illustrated in Fig. 36). In principle, the advantages stated for Fourier spectroscopy are applicable also for Hadamard transform spectroscopy which employs a conventional grating spectrometer. As our consideration are mainly devoted to Fourier spectroscopy, there is not sufficient space to present and discuss here all the details about realisation of Hadamard spectroscopy, *i.e.* experimental difficulties, limits of the method (diffraction at the slots!) etc. Clearly, a Hadamard spectrometer is superior to a conventional scanning grating spectrometer. In principle, the performance of a Hadamard spectrometer is similar to that of a Fourier spectrometer [79]. The result of a comparison of actual instruments of the two kinds will depend on the technical details of spectrometer design. In practice, it can be of some advantage to employ a Hadamard spectrometer and not a Fourier spectrometer when only a narrow spectral range is of interest. A Hadamard spectrometer contains still a grating monochromator, and a narrow spectral range can easily be selected for which the $N$ Hadamard signals are then recorded. For Fourier spectroscopy, narrowing of the spectral range means decreasing the signal at the detector but not reducing the work necessary to scan an interferogram. Another point worth mentioning is that mechanical precision and tolerances are less for Hadamard spectrometers in comparison to Michelson interferometers. These apsects will be discussed in more detail in the next section. As well as Fourier spectroscopy, Hadamard spectro-. scopy has been employed for special applications in space and in astronomy [80,81]

The critical comparison with other methods has already led us to some disadvantages of Fourier spectroscopy which we have to consider in a systematic way now, after commenting extensively on the advantages. Many people well acquainted with the practice of spectroscopy have regarded it as a major disadvantage of this method that a computer is necessary for the execution of the Fourier transform. In fact, it was troublesome to scan the interferogram, to

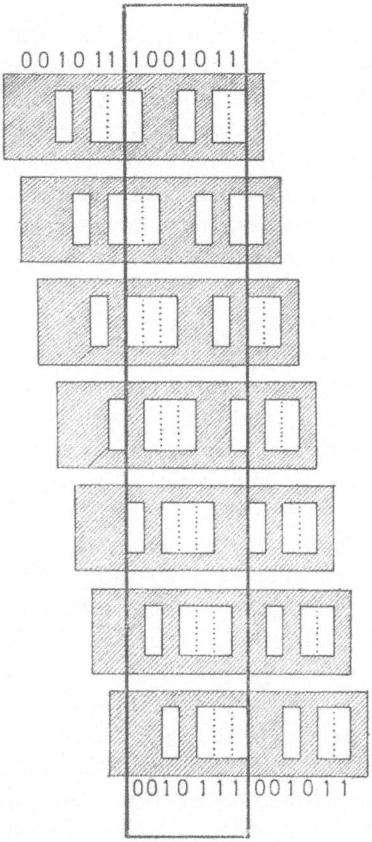


Fig. 37. Hadamard mask with a total of 13 slots for demonstration of the principle. Only 7 slots (within the section marked by the heavy line) are exposed to the radiation. 7 independent measurements are obtained by shifting the mask stepwise, each step corresponding to the width of 1 slot on the mask. The resulting Hadamard encoding matrix is a 7 × 7 cyclic matrix

take the data to the computer facility, and to have to wait for the final results. And not every spectroscopist had easily access to a computer in these early days of Fourier spectroscopy. With the development of relatively cheap, small digital computers, these handicaps have been eliminated from Fourier transform spectroscopy. All commercial instruments are provided with computers, and their operation is not more difficult than that of a conventional two-beam spectrometer. Another objection is that the recorded interferogram contains all the information but is not intelligible to the spectrocopist. But this handicap also has been overcome by modern technique with the rapid scan method in the near- and middle-infrared where the spectral data are available very soon. For the slow-scan spectroscopy in the far infrared, as was already mentioned, the real-time Fourier transform is a mean to visualize all information in the familiar form of a spectrum while the interferogram is still being recorded. Of course, the disadvantage of the real-time method is that only single-sided interferograms can be used where the spectra are very sensitive to phase errors (cf. Section 5.3). And if the normal integration method (see Section 4.3) is applied in case of a slow-scan instrument, the coarse contours of the spectrum can be read from the interferogram with some experience. To be fair, let us note that the necessity of Fourier transform and of using a computer brings also advantages. For example, the spectral window function can be taylored as desired by the choice of an appropiate apodization. Secondly, amplitude Fourier transform spectroscopy offers the possibility of measuring the magnitude and the phase of a reflection or transmission coefficient. And finally, the computer can be used for all data processing necessary to obtain the final results in every kind of investigation, as already mentioned.

When considering the advantages and disadvantages of spectroscopic methods, it is regarded an advantage of Fourier spectroscopy to cover a wide spectral range without difficulty. In the far infrared range, this is a true advantage without any seamy side. In the middle- and near-infrared however, problems can arise. One of the essential properties of Fourier spectroscopy is that the signal at the detector is increased as the spectral range is widened. This increase in signal improves the signal-to-noise ratio as long as the noise is independent of the signal. If the signal becomes too large, the detector may show saturation effects and nonlinearities. Generally these effects cause a more or less drastic flattening of the grand maximum of the interferogram. Now, all our considerations depend on the assumption that the electrical signal obtained in scanning the interferogram is proportional to the intensity of light. If this linearity is destroyed by saturation effects of the detector, aggravating errors will arise in the Fourier transformed spectral data. Thus, the strength of the signal must not exceed the limits of the dynamical range of the detector and of the electronic system. There are more problems that may originate from the fact that a Michelson interferometer is not a monochromator and that it transmits the total intensity of radiation. If, for example, samples cooled to rather low temperatures are investigated, the total radiation from a wide spectral range could cause considerable heating of the sample by absorption processes. Again, this will be more a handicap in the middle- and near-infrared and not so much in the far infrared. In the case of such a handicap, it can be necessary to waive partly the multiplex gain and reduce the light intensity on the sample and, as a consequence, the absorption of light and the heating.

In the visible, Fourier spectroscopy is rather trickly and not so advantageous. A high mechanical precision and a very good alignment of the instrument are required for these short waves. Moreover, the multiplex gain partly breakes down due to the fact that the dominating noise is no longer detector noise but photon noise which varies with the strength of the signal (cf. Section 5.4). On the whole, the domain of Fourier spectroscopy is the infrared spectral range where the advantages can be fully realized and where the disadvantages mentioned can be limited to a tolerable magnitude.

## 5.2 Precision, Alignment, Filtering

The required mechanical precision, the perfection of the alignment, and the accuracy needed in the interferogram are not basic principles but rather technical aspects of Fourier transform spectroscopy. However, their importance must not be underestimated in the everyday life of a spectroscopist. One essential difference between the Michelson interferometer and the grating spectrometer is that in the latter case the required mechanical precision is more or less concentrated in the grating. The relatively high cost of the grating is due to this precision. The basic, optical equipment is much cheaper for the Michelson interferometer, but the alignment has to be more precise in the actual interferometer otherwise the oscillatory part of the interferogram will be reduced considerably and the interference spoilt. The two mirrors or their images have to be kept parallel to within a tenth of the smallest wavelength in the spectrum, *i.e.* 1 $\mu$m for the range 10 to 1000 $cm^{-1}$. This involves the adjustment of the mirrors and of the beam divider as well as the mechanics of the carriage on which the movable mirror is mounted. In the case of a conventional spectrometer, the absolute measurement of a wavelength and its accuracy depend on the grating constant and therefore on the precision of the grating. For a Michelson interferometer, they depend on the Moiré or laser system by means of which the path of the movable mirror and thus the path difference is controlled. Typical figures for the sampling interval $\Delta s$ have been quoted in Section 3.2, *e.g.* $\Delta s = 5$ µm for the range 10 to 1000 $cm^{-1}$. The error for the determination of the path travelled by the movable mirror should be smaller than $1/10\,\Delta s$. From these figures, it is clear that these requirements are met most easily in the far-infrared spectral region with wavelengths of about 100 $\mu$m. On the other hand, this is the region where the energy limitations are most severe. These are the two essential reasons why Fourier transform spectroscopy is the preferred method in the far infrared. But as already mentioned, the precision requirements can be met also in the near- and middle-infrared region. Especially with respect to this region, the arguments concerning mechanical precision and costs of the basic equipment can be extended to Hadamard spectroscopy. Here again, the precision is bought for a relatively high price with the grating. And for a realistic comparison, also these arguments have to taken into account, in addition to the advantages and disadvantages discussed in the preceding section. Moreover, the spacing of the grooves of a grating may change with temperature while the wavelength precision of a laser controlled interferometer is independent of such influences. For special applications, for example spectroscopic investigations in space and astronomy [82–85], fourier spectroscopy was preferred for reasons of the lower cost, of the smaller size and of the lower weight of the interferometer. Even under the extreme conditions during space vehicle reentry into the earth atmosphere, interferograms were recorded by a Fourier spectrometer [86]. Moreover, the interferogram data can be taken with a small interferometer, and the Fourier transform may be executed elsewhere.

As regards the throughput advantage (see Section 5.1), the power flux through a grating spectrometer and a Michelson interferometer have been compared. A more realistic approach would have to include the reflection and absorption losses in the filters for the supression of unwanted radiation. In grating spectrometers, the bandwidth of the radiation has to be reduced to one octave or less. The losses due to filters generally amount to more than 50 % of the wanted radiation. As the filter problems are greatly reduced in Fourier spectroscopy, losses due to filters are much less. Adding this advantage to the other two already discussed ("throughput" and "multiplex"), we see that the signal-to-noise ratio will be better by several orders of magnitude than in a grating spectrometer. However, this does not neces sarily mean that the errors and the noise in the spectrum are reduced by the same factor. In other words, the excellent overall signal-to-noise ratio of the interferogram may conceal the fact that the signal-to-noise ratio is less than unity for some finer details in the spectrum.

In order to understand this important property of Fourier transform spectroscopy, let us consider a broad spectrum over a wide wave number range with one narrow absorption line in it (cf. Fig. 12). Then, according to the rules of Fourier transformation, the broad spectrum produces an interferogram with highly damped oscillation. The maximum amplitude of the oscillation and also the mean value of the interferogram are equal to the total intensity or to the area under the spectral distribution [see Appdx 1 and Eqs. (A 1.1) and (A 1.2)]:

$$\int_0^\infty I_{\text{Broad}}(\tilde{\nu})\,d\tilde{\nu} \approx f_{\text{Broad}} \cdot \gamma_{\text{Broad}}\,, \tag{5.15}$$

where $f_{\text{Broad}}$ and $\gamma_{\text{Broad}}$ are the maximum value and the width of the broad spectrum, respectively. The narrow line causes a less damped oscillation in the interferogram (see Fig. 12) and its maximum amplitude is equal to the area covered by the narrow line [see Eqs. (A 1.1) and (A 1.2)]:

$$\int_0^\infty I_{\text{Narrow}}(\tilde{\nu})\,d\tilde{\nu} \approx f_{\text{Narrow}} \cdot \gamma_{\text{Narrow}}\,, \tag{5.16}$$

where $f_{\text{Narrow}}$ and $\gamma_{\text{Narrow}}$ are the maximum value and the width of the narrow line, respectively. If the Fourier transform effected by the computer is to yield the narrow line without distortion in the spectrum, the corresponding oscillation in the interferogram must not be obscured by noise. The least required for this purpose is that the RMS value of the noise be equal to the amplitude of the narrow line [see Eq. (5.16)]. Then the signal-to-noise ratio is equal to unity for this line while it is much greater for the broad spectrum or for the interferogram as a whole. On these grounds, a quality factor $Q$ is defined [65]:

$$Q = \frac{f_{\text{Broad}} \cdot \gamma_{\text{Broad}}}{f_{\text{Narrow}} \cdot \gamma_{\text{Narrow}}}\,, \tag{5.17}$$

where $f_{\text{Broad}} \cdot \gamma_{\text{Broad}}$ is a measure of the total intensity (area under the spectral distribution), and $f_{\text{Narrow}} \cdot \gamma_{\text{Narrow}}$ a measure of that of the finest detail in the spectrum. The signal-to-noise ratio of the interferogram must be better than or at least equal to the quality factor $Q$. Conversely, the effective quality factor $Q$ by which the finest measurable details in the spectrum are defined must be less than or equal to the signal-to-noise ratio.

### 5.3 Possible Errors and their Correction

In this section, attention is focused on possible errors and how to avoid or correct them. There are a number of possibilities of obtaining erroneous spectra in Fourier transform spectroscopy [66], and it is useful to be aware of them since often slight errors in the interferogram will have a strong effect on the computed spectrum because it is not the error itself but its Fourier transform that affects the spectrum. This is a typical situation in Fourier transform spectroscopy and we have to take it into account. Generally the effect of errors is less serious when a double-sided interferogram rather than a single-sided interferogram is used to compute the spectrum. Fig. 38 demonstrates the increase in uncertainty for the single-sided interferogram in comparison to the double-sided interferogram, especially at the ends of the spectral range where the signal-to-noise ratio is rather low. In the ideal case without any errors, we would subject a single-sided interferogram to a cosine transform, and in most practical cases, the Fourier transform is executed this way. But, as the examples show, we have to be especially careful about errors in this case of a single-sided interferogram. In the case of a double-sided interferogram, we have to pay for better accuracy by processing twice as many data and taking more computer time. Without any errors we would need only the cosine transform for the double-sided interferogram. But in practice, the sine transform is also performed to eliminate some of the errors and their effects.

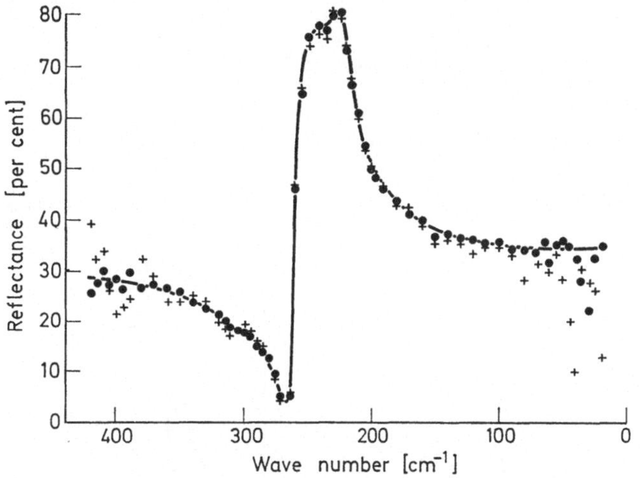


Fig. 38. Reflectance of $Mg_2$ Ge computed from single-sided (+) and double-sided (•) interferograms, according to Eq. (5.20) in the latter case (taken from Ref. [42])

In a systematic way, we distinguish intensity errors and phase errors in the interferogram. For example, intensity errors arise when an incorrect mean value has been subtracted from the interferogram, or when the mean value is changing due to possible temperature-dependent change in the gain of an amplifier. Also the errors due to nonlinearities in the detector or the electronic system are intensity errors. A last sort of intensity error will result in the computed spectrum if an erroneous electrical puls is recorded together with the interferogram. A phase error arises when the position $s = 0$ has not been determined with sufficient ac-

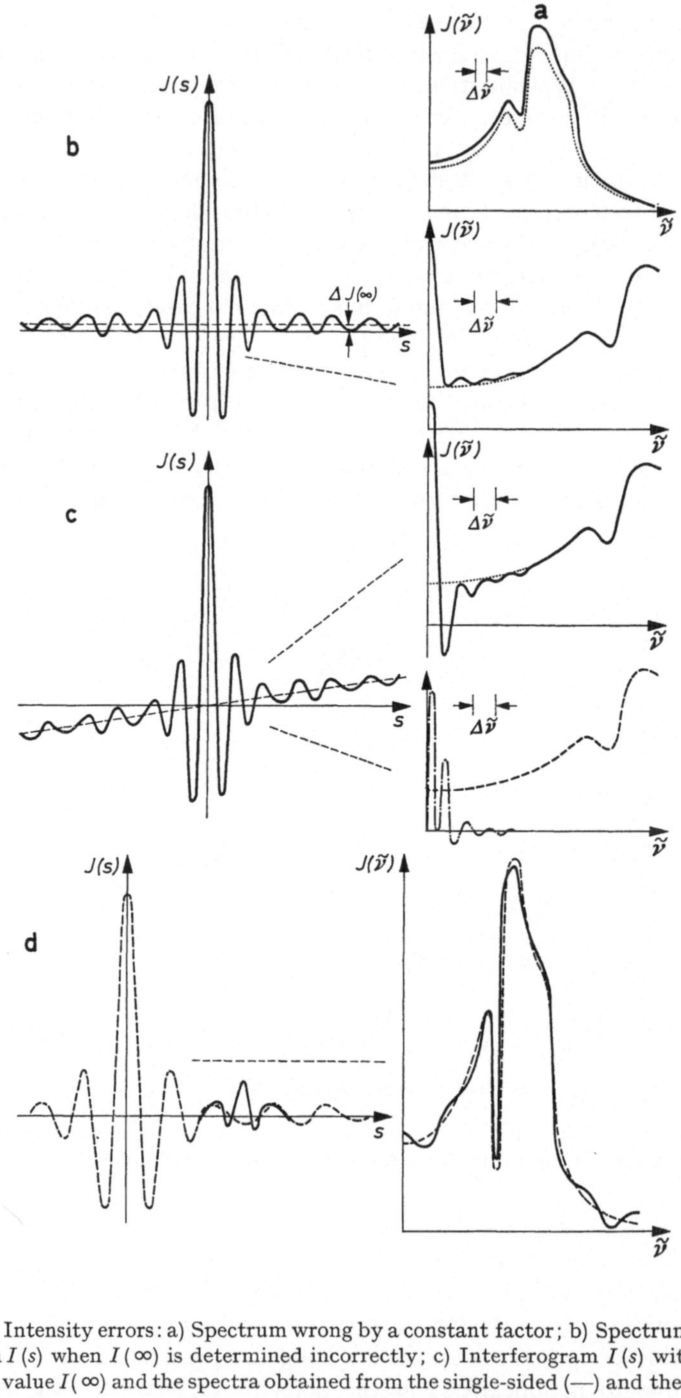


Fig. 39 a-d. Intensity errors: a) Spectrum wrong by a constant factor; b) Spectrum $I(\tilde{\nu})$ and interferogram $I(s)$ when $I(\infty)$ is determined incorrectly; c) Interferogram $I(s)$ with linear drift of the mean value $I(\infty)$ and the spectra obtained from the single-sided (—) and the double-sided interferogram (- - - = cos transform, — · — = sine transform). The dotted line (. . .) indicates the undistorted spectrum. The spectrum obtained by cos transform (- - -) shows no deviation from the undistorted spectrum; d) Interferogram $I(s)$ (- - -) with an erroneous impulse (——) and corresponding spectrum (——) as well as the undistorted spectrum (- - -)

curacy. This can easily occur, for example, when the Fourier transform is performed in the digital way. Then, there may be no sample point taken exactly at $s = 0$. Misalignment of the Michelson interferometer may cause an asymmetrically distorted interferogram [87,88]. As will be explained later, this case corresponds to a nonlinear phase error. The compilation and discussion of errors serves mainly the purpose to introduce the reader to the errors and their effect on the computed spectrum. It is certainly advisable to know about the possibilities of errors and their signature in the interferogram. Those who are aware of the errors from the interferogram will interpret the spectra with sufficient caution and will try to eliminate them. The other purpose of this section is to show the principles of the methods used for the correction of phase errors in commercially available instruments. It seems advisable to concentrate on the systematic errors in this section and their possible correction. The statistic errors in an interferogram will be discussed in the next section in context with the noise problems.

Let us now turn to the intensity errors and discuss the most frequent ones in detail: Figs. 39a)–d) demonstrate some of them in the interferogram and the effect in the spectrum. A minor intensity error is that sometimes the computed spectrum happens to be wrong by a constant factor due to some change in the amplifier gain between background and sample measurement (Fig. 39a). This does not affect the structures in the spectrum but is important when refractive index or absorption coefficient are evaluated from the reflectance or transmittance, respectively.

In computing the spectrum, the mean value has to be subtracted from the recorded interferogram, except for some special cases like phase-modulation Fourier transform spectroscopy. Here errors can arise when the mean value $I(\infty)$ is not accurately determined (see Fig. 39b) or changes during the scan due to a drift of the mean value (Fig. 39c). In both cases, the effect on the spectrum is obtained by a Fourier transform of the mathematical form of these errors, which are slowly varying functions of the path difference $s$ as against the interference patterns of the interferogram. According to the elementary rules of Fourier transforms, these error functions produce erroneous structures of the spectrum in the neighborhood of $\tilde{\nu} = 0$. For the offset of $I(\infty)$, no difference is encountered, whether a single-sided or double-sided interferogram is used to evaluate the spectrum. For the drift of $I(\infty)$, however, a difference is exhibited. In all these cases, the erroneous structures are important only in a wave number range extending to a few multiples of $\Delta\tilde{\nu} = \frac{1}{s_{max}}$ from $\tilde{\nu} = 0$. Consequently, this spectral range should be rejected if strange and unexpected structures appear here in the spectrum. Another intensity error occurs when an extra impulse is produced in the interferogram (Fig. 39d), perhaps due to switching some apparatus off or on. If the impulse is sufficiently short in time (proportional to path difference $s$), the Fourier transform yields a nearly undamped cosine wave superimposed on the spectrum. The remedy recommended in this case is to replace the short impulse in the interferogram by a smooth function and in this way to remove the cause of the cosine wave in the spectrum. As already mentioned, distortions in the spectra due to nonlinearities and saturation effects are also intensity errors. The experimenter may recognize them from a flattening of the grand maximum of the interferogram. In most practical cases, these errors can only be suppressed by narrowing the spectral range and thus reducing the signal (cf. discussion in Section 5.1). And even for the spectroscopy in the middle- and far-infrared spectral region, it may prove very useful to check the linearity of the system in order to avoid uncontrolable distortions in the Fourier transformed spectra.

Now, let us consider phase errors. As already pointed out, an error arises when the true origin of the interferogram is missed by a small path difference $\varepsilon < \Delta s$ (Fig. 40a) where $\Delta s$ is the sampling interval. This error is called a linear phase error because $2\pi\tilde{\nu}\varepsilon$ means an erroneous phase shift in the interferogram function, which is linear with respect to the wave number $\tilde{\nu}$. Including the effects of truncation and apodization, we obtain for the cosine transform of the double-sided interferogram with a phase error $\varepsilon$ approximately [68,69,70]:

$$I^{S}_{\text{obs}}(\tilde{\nu}) = \int_{-s_{\max}}^{+s_{\max}} \bar{I}(s+\varepsilon)\, S(s) \cos(2\pi\tilde{\nu}s)\, ds \approx \cos(2\pi\tilde{\nu}\varepsilon) \cdot I_{\text{obs}}(\tilde{\nu}) \qquad (5.18)$$

where $S(s)$ is the apodization function and $I_{\text{obs}}(\tilde{\nu})$ the spectrum we would obtain for $\varepsilon = 0$ [cf. Eqs. (3.6) and (3.7)]. For the sine transform, we obtain for the double-sided interferogram approximately

$$I^{A}_{\text{obs}}(\tilde{\nu}) = -\int_{-s_{\max}}^{+s_{\max}} \bar{I}(s+\varepsilon)\, S(s) \sin(2\pi\tilde{\nu}s)\, ds \approx \sin(2\pi\tilde{\nu}\varepsilon) \cdot I_{\text{obs}}(\tilde{\nu})\,. \qquad (5.19)$$

For $\varepsilon = 0$, $I^{A}_{\text{obs}}(\tilde{\nu})$ would be zero. From Eqs. (5.18) and (5.19), a good approximation for $I_{\text{obs}}(\tilde{\nu})$ (without phase error) is obtained by means of the following relation

$$I_{\text{obs}}(\tilde{\nu}) \approx \sqrt{[I^{A}_{\text{obs}}(\tilde{\nu})]^2 + [I^{S}_{\text{obs}}(\tilde{\nu})]^2}\,. \qquad (5.20)$$

The accuracy and the validity of the approximation used in Eqs. (5.18—5.20) depends on two presumptions: At first $\varepsilon$ has to be small, and in practice, $\varepsilon$ will not exceed $\frac{1}{2}\Delta s$. Secondly, the width of the central maximum of the spectral window function has to be sufficiently small that the variation of $\cos(2\pi\tilde{\nu}\varepsilon)$ and $\sin(2\pi\tilde{\nu}\varepsilon)$ can be neglected within this width.

For the case of a single-sided interferogram, the effect of the phase error on the computed spectrum is more clamaging. As Fig. 40a shows, the phase error causes asymmetric distortions in the spectrum, and there is no simple way of removing the effect of the error from the spectrum. Thus, we have to correct the single-sided interferogram for the phase error. For reasons of economy, we would like to scan a single-sided interferogram only. And it was already mentioned that scanning a single-sided interferogram is the only possibility for real-time Fourier analysis. In practice, there are several methods for the correction of phase errors. The methods which are suitable for both, linear and nonlinear phase errors, will be discussed later. Here we will concentrate on linear phase errors. One way often employed to eliminate this phase error is the so-called parabola fit. For this, we consider the three digital points of the interferogram closest to the

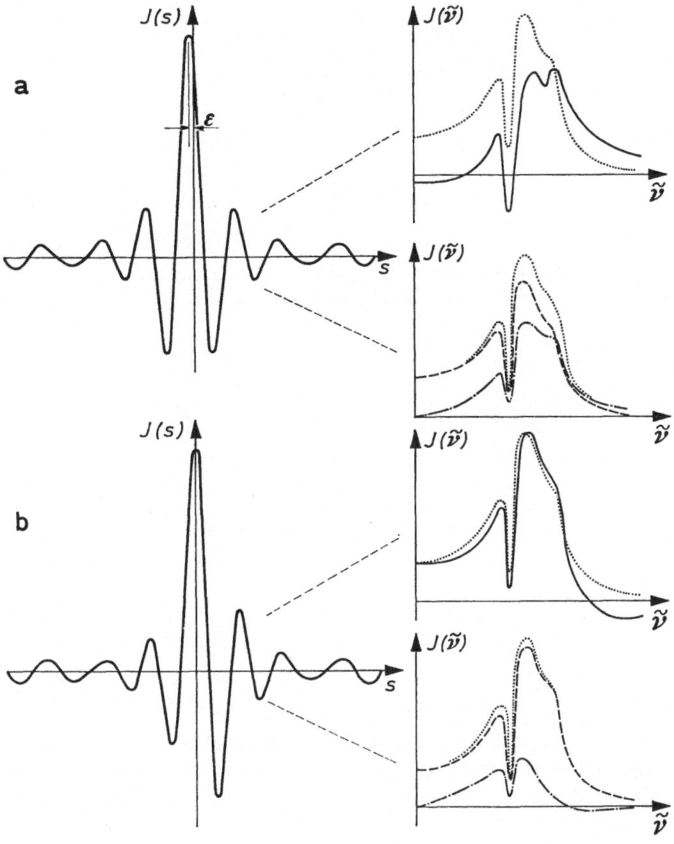


Fig. 40. Phase errors: a) linear phase error; b) nonlinear phase error. — Interferograms $I(s)$ and spectra $I(\tilde{\nu})$ obtained from a single-sided interferogram (—) and from a double sided one (- - - = cos transform, —·— = sine transform). — The dotted line (. . .) indicates the undistorted spectrum which would have been obtained without any errors

true origin $s = 0$. On the basis of these three points, a parabola is constructed as an approximation to the interferogram function. The maximum of the parabola is usually sufficiently close to $s = 0$ to exclude phase error effects, and this maximum is used for $s = 0$ in order to avoid or correct the phase error in the interferogram.

When the Michelson interferometer with finite aperture is not properly adjusted nonlinear phase errors arise [87]. These phase errors are no longer linearly dependent on the wave number $\tilde{\nu}$, and they cause an asymmetric distortion of the interferogram (Figs. 40b and 41). It should be noted that all illustrations in connection with errors (Figs. 39, 40 and 41) have been produced by computer simulation (cf. Appendix 1). In order to make the essential features as clear as possible the effects of finite resolution etc. are left out where they have not necessarily to be included. In these cases, the resolution width $\Delta\tilde{\nu}$ is given in the figure (Figs. 39a—c). In Fig. 41, the error correction is demonstrated with finite

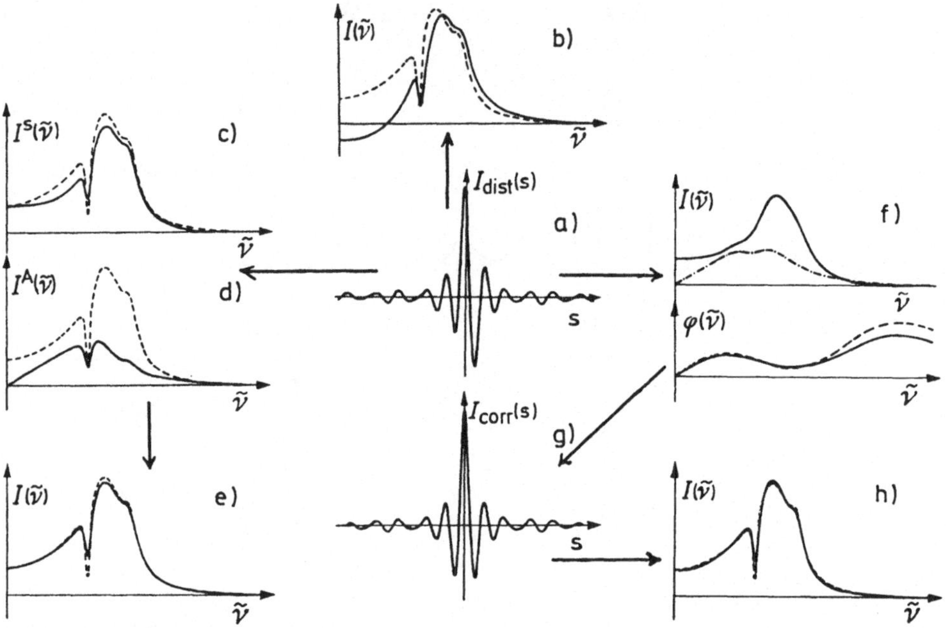


Fig. 41. Phase error correction:

a) distorted interferogram $I_{dist}(s)$, with phase errors.

b) spectrum $I(\tilde{\nu})$ (—) calculated by means of cos-transform from the half of $I_{dist}(s)$ with $s \geq 0$ (single-sided distorted interferogram).

c), d) spectra $I^S(\tilde{\nu})$ and $I^A(\tilde{\nu})$ (—) calculated by means of cos- and sin-transform, resp., from $I_{dist}(s)$ (double-sided distorted interferogram).

e) corrected spectrum $I(\tilde{\nu}) = \sqrt{(I^S)^2 + (I^A)^2}$ (—) as obtained from the double-sided distorted interferogram using the results c) and d).

f) spectra $I(\tilde{\nu})$ calculated by means of cos-transform (—) and sin-transform (—·—) from a small double-sided portion of $I_{dist}(s)$ and phase error $\varphi(\tilde{\nu})$ evaluated from these two spectra.

g) symmetric interferogram $I_{corr}(s)$ corrected for phase errors, evaluated from $I_{dist}(s)$ by means of $\varphi(\tilde{\nu})$ (see part f)).

h) corrected spectrum cos-transform from the half of $I_{corr}(s)$ for $s \geq 0$ (single-sided corrected interferogram).

For comparison, the true spectrum (- - -) is given in b), c), d), and h) and the true phase error (- - -) in f) where "true" means as would have been obtained with infinite resolution

resolution and with apodization. Let us return to the problem of a general phase errors $\varphi(\tilde{\nu})$. In the interferogram, $\varphi(\tilde{\nu})$ means a wave number dependent phase shift of the contributions $I(\tilde{\nu})$ at wave number $\tilde{\nu}$

$$I_{dist}(s) = 2 \int_0^\infty I(\tilde{\nu}) \left[1 + \cos(2\pi\tilde{\nu}s + \varphi(\tilde{\nu}))\right] d\tilde{\nu} . \tag{5.21}$$

In fact, it is this phase shift $\varphi(\tilde{\nu})$ which causes the asymmetric distortion of the interferogram. For $\varphi(\tilde{\nu})$, linear or nonlinear function of wave number $\tilde{\nu}$, the assumption generally holds that it is a smooth and slowly varying function of $\tilde{\nu}$.

With almost the same approximation as used for deriving of Eqs. (5.18) and (5.19), the cosine and sine transforms yield for a double-sided interferogram

$$
\begin{aligned}
I^{\mathrm{S}}_{\mathrm{obs}}(\tilde{\nu}) &\approx \cos\varphi(\tilde{\nu}) \cdot I_{\mathrm{obs}}(\tilde{\nu}) \\
I^{\mathrm{A}}_{\mathrm{obs}}(\tilde{\nu}) &\approx \sin\varphi(\tilde{\nu}) \cdot I_{\mathrm{obs}}(\tilde{\nu}) \,.
\end{aligned}
\tag{5.22}
$$

Again, the root of the sum of the squares of $I^{S}_{\mathrm{obs}}(\tilde{\nu})$ and $I^{A}_{\mathrm{obs}}(\nu)$ is a good approximation for $I_{\mathrm{obs}}(\tilde{\nu})$ unless the interferogram is too badly distorted (cf. Fig. 41). Also for the nonlinear phase error, the cosine transform of a single-sided interferogram yields untolerable distortions of the spectrum (cf. Figs. 40a and 41). An error correction is necessary to obtain a reliable spectrum. This can be done only with the knowledge of $\varphi(\tilde{\nu})$. As $\varphi(\tilde{\nu})$ is a smooth and slowly varying function of $\tilde{\nu}$, only a low resolution and, correspondingly, only a small double-sided portion of the interferogram is needed for the determination of $\varphi(\tilde{\nu})$ (cf. Fig. 41).

$$
\varphi(\tilde{\nu}) = \operatorname{arc\,tg} \frac{I^{A}_{\mathrm{obs}}(\tilde{\nu})}{I^{S}_{\mathrm{obs}}(\tilde{\nu})} \tag{5.23}
$$

In other words, the correction of a phase error $\varphi$ requires a short double-sided interferogram around $s=0$ regardless whether the whole interferogram is recorded single-sided or double-sided. The mostly used correction method was first proposed by M. Forman [68,70]. After determining $\varphi(\tilde{\nu})$, the next step of this method is to calculate what can be called the Fourier transform of $e^{i\varphi(\tilde{\nu})}$

$$
F(s) = 2 \int_{0}^{\infty} [\cos\varphi(\tilde{\nu}) \cos(2\pi\tilde{\nu}s) + \sin\varphi(\tilde{\nu}) \sin(2\pi\tilde{\nu}s)] \, d\tilde{\nu} \tag{5.24}
$$

With the help of this function $F(s)$, the signatures of the error are removed from the distorted interferogram $I_{\mathrm{dist}}(s)$ and a symmetric interferogram $I_{\mathrm{corr}}(s)$ is obtained (cf. Fig. 41) by means of the following integration:

$$
\bar{I}_{\mathrm{corr}}(s) = \int \bar{I}_{\mathrm{dist}}(s') \cdot F(s-s') \, ds' \tag{5.25}
$$

where $\bar{I}_{\mathrm{corr}}(s)$ and $\bar{I}_{\mathrm{dist}}(s)$ are only the oscillatory parts of the corrected and distorted interferogram, respectively. It can be verified that $I_{\mathrm{corr}}(s)$ is the interferogram that would have been recorded without any phase error. We insert the Fourier transforms of $I_{\mathrm{dist}}(s')$ [cf. Eq. (5.21)] and of $F(s)$ [see Eq. (5.24)] into Eq. (5.25). Then we arrive at the approximate result

$$
\bar{I}_{\mathrm{corr}}(s) \approx 2 \int_{0}^{\infty} I(\tilde{\nu}) \cos(2\pi\tilde{\nu}s) \, d\tilde{\nu} \tag{5.26}
$$

which is just the oscillatory part of the interferogram $I(s)$ without phase error [cf. Eqs. (3.2) and (3.5)]. Finally, the Fourier transform of $I_{\mathrm{corr}}(s)$ yields the desired spectrum $I(\tilde{\nu})$ (cf. Fig. 41).

This method of error correction can be used for single-sided as well as for double-sided interferograms. In case of a single-sided one, it is the only mean to deduce the spectrum $I(\tilde{\nu})$ from the experimental data $I_{dist}(s)$. And in case of a double-sided one, it can be more economic to determine $\varphi(\tilde{\nu})$ by means of cosine and sine transform for a small portion, to correct the whole interferogram and to calculate $I(\tilde{\nu})$ only be means of the cosine transform. For most routine spectroscopic investigations, the simplest and most economic way is to scan a small double-sided interferogram and to extend it single-sided to maximum path-difference. Then, the phase errors which are nearly unavoidable in the near- and middle-infrared can be corrected. And the computer program for most commercial instruments in that range has a subprogram to correct the usually single-sided interferogram for the phase errors by means of the method due to M. Forman [68,70] which was described in detail here. Of course, this error correction is not exact in a mathematical sense but a good approximation since the integrals in Eqs. (5.24) and (5.25) cannot be extended to infinity and have to be truncated. However, an other method proposed by L. Mertz [69] for phase error correction was shown to be less efficient [70] though improvements were considered recently [89].

At last it remains to point out that phase errors are not only unwanted and under certain conditions unavoidable phenomena in Fourier spectroscopy the elimination of which is a necessary but rather involved and elaborate procedure. In contrast, there are special applications of Fourier spectroscopy where phase errors have been introduced deliberately. In this method which is known as "chirping" [86,90–92], a plane parallel plate called chirping plate is placed in one arm of the Michelson interferometer or two different plates are placed one in each arm. These plates introduce a wave number dependent phase shift

$$\varphi(\tilde{\nu}) = 2\pi\tilde{\nu}\,[n(\tilde{\nu}) - 1]\,d \tag{5.27}$$

as we learnt in Section 4.7 about amplitude Fourier spectroscopy [cf. Eq. (4.18)]. For two plates, the resultant phase shift is the difference of two expressions like the one given in Eq. (5.27). If the spectrum under investigation consists for example of two clearly separated emission bands, the effect of chirping on the interferogram is that the central fringes corresponding to the two emission bands are shifted away from $s=0$ to different path differences $s_1=[n(\tilde{\nu}_1)-1]d$ and $s_2=[n(\tilde{\nu}_2)-1]d$ according to the variation of the refractive index $n(\tilde{\nu})$ of the chirping plate between the two bands centered at wave numbers $\tilde{\nu}_1$ and $\tilde{\nu}_2$ (cf. Fig. 42). That means that the large central maximum of the interferogram disappears. Its intensity is shifted away from $s=0$ and is distributed over a certain path difference range. As it is relatively easy to correct the spectra for the phase error in this case ("dechirping"), this method offers some advantages with respect to the "dynamic range problem". At first, there is no extremely high peak intensity at $s=0$. Secondly, there is a possibility to correct for intensity errors due to nonlinearities in this case [91]. For these reasons, the spectra obtained so far by the use of chirping in Fourier spectroscopy are in the near-infrared range (2000–6500 cm$^{-1}$). The interferograms for the presented data were recorded on rocket-borne instruments [86,92].

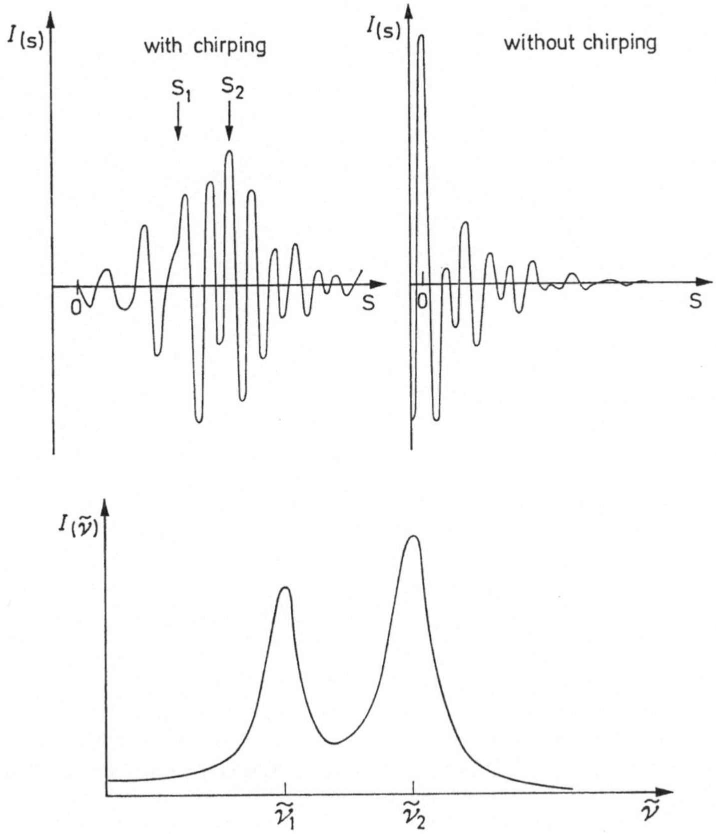


Fig. 42. Interferogram $I(s)$ obtained with and without chirping and corresponding spectrum $I(\tilde{\nu})$ with two bands centered at $\tilde{\nu}_1$ and $\tilde{\nu}_2$. In the case of chirping, the intensity of the grand maximum of the interferogram is shifted away from $s=0$ to different path differences $s_1 = [n(\tilde{\nu}_1)-1]d$ and $s_2=[n(\tilde{\nu}_2)-1]d$ as indicated

## 5.4 Noise Problems

In considering noise problems in Fourier transform spectroscopy, we realize that what was found to be a property of Fourier transform spectroscopy with respect to systematic errors is also true for the statistical errors or noise. The errors arise in the interferogram, but we want to know their effect on the spectrum. Mathematically, the connection is given by the Fourier transform. Before treating the noise problems in detail, we have to distinguish several kinds of statistical errors in Fourier spectroscopy. At first there are fluctuations $N(s)$ in the interferogram (Fig. 43). These fluctuations can be due to detector noise and noise in the electronic system. Then they are independent of the signal $I(s)$. And they can be fluctuations of the signal itself. Among these, source fluctuations of a mercury arc have been shown to add perceptably to the noise in the far-infrared when the interferogram is recorded with a rather low speed [92]. In the visible region, photon

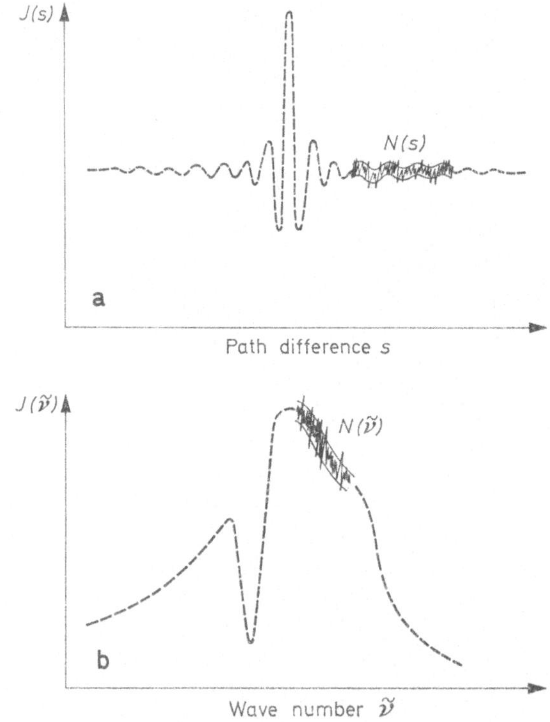


Fig. 43. Noise amplitude $N(s)$ in the interferogram (a) and noise amplitude $N(\tilde{\nu})$ in the spectrum (b)

noise has to be taken into account. In this case, the profit from the advantages of Fourier spectroscopy is reduced [94,95]. In infrared spectroscopy, detector noise usually dominates. For the following considerations therefore, let us assume that $N(s)$ is of this kind and is independent of the signal $I(s)$. In addition, statistical errors can arise when the interferogram is sampled for the digital Fourier transform. Such sampling errors will be discussed at the end of the section.

Now let us concentrate on the properties of the noise amplitude $N(s)$ under the assumptions made above. The aim of these considerations is to derive some realistic expressions for the signal-to-noise ratio in infrared spectroscopy and its dependence on experimental parameters like scanning time, resolution etc. Since $N(s)$ is a statistical function, its average $N(s)$ will be zero. With the computation of the spectrum, the noise $N(s)$ is also subjected to multiplication by the scanning function $S(s)$ and to the Fourier transform. The result is the noise amplitude in the spectrum (Fig. 43)

$$N(\tilde{\nu}) = \int_{-s_{\max}}^{+s_{\max}} N(s) \cdot S(s) \cdot \cos(2\pi\tilde{\nu}s)\, ds \,. \tag{5.28}$$

For $N(\tilde{\nu})$ also, the average $\overline{N(\tilde{\nu})}$ will be zero. We recall that for noise problems the essential and reproducible quantity is the root mean square (RMS) value. For the interferogram and for the spectrum, the RMS values of the noise may be defined as follows:

$$N_s = \sqrt{\overline{N^2(s) \cdot S^2(s)}} = \sqrt{\frac{1}{2s_{max}} \int_{-s_{max}}^{+s_{max}} N^2(s) S^2(s)\, ds}$$
$$N_{\tilde{\nu}} = \sqrt{\overline{N^2(\tilde{\nu})}} = \sqrt{\frac{1}{\tilde{\nu}_{max}} \int_0^{\tilde{\nu}_{max}} N^2(\tilde{\nu})\, d\tilde{\nu}}\,. \tag{5.29}$$

This definition takes into account that the interferogram with its noise is multiplied by $S(s)$ before the Fourier transform is carried out, and that the effective noise amplitude is $N(s) \cdot S(s)$ instead of $N(s)$. Furthermore, it is presumed that the bandwidth $\Delta f$ of the electronics is selected to be appropiate for the maximum wave number $\tilde{\nu}_{max}$. We recall from our considerations in Section 4.6 that the time constant $\tau$ limiting the bandwidth of the electronic system should have a value close to (cf. Section 4.6)

$$\tau = \frac{1}{3} \frac{\Delta s}{v} = \frac{1}{6v\tilde{\nu}_{max}}$$

where $\Delta s$, $v$, $\tilde{\nu}_{max}$ are the sampling interval, the speed of the movable mirror, and the maximum wave number in the spectrum, respectively. We recall further that each wave number $\tilde{\nu}$ corresponds to a signal frequency $f = 2v\tilde{\nu}$ [cf. Eq. (4.3)] and that the maximum frequency to be transmitted by the electronic system is $f_{max} = 2v\tilde{\nu}_{max}$. For a proper choice of $\tau$ therefore, the bandwidth $\Delta f \sim \frac{1}{\tau}$ of the electronics is proportional to $\tilde{\nu}_{max}$

$$\Delta f \sim 2\, v\, \tilde{\nu}_{max}\,. \tag{5.30}$$

Now, the spectrum of noise frequencies is not necessarily limited to the range $0 \leqq f \leqq f_{max}$. But the frequency components for $f > f_{max}$ will be considerably attenuated. As each frequency in the electrical signal corresponds to a wave number ($f = 2v\tilde{\nu}$!), the Fourier transform $N(s) \rightarrow N(\tilde{\nu})$ yields essentially the spectrum of noise frequencies. And from our considerations, $N(\tilde{\nu})$ will be rather small for $\tilde{\nu} > \tilde{\nu}_{max}$ corresponding to $f > f_{max}$. On these grounds, we can neglect $N(\tilde{\nu}) \approx 0$ above $\tilde{\nu} = \tilde{\nu}_{max}$ and limit the integration over $\tilde{\nu}$ to the range $0 \leqq \tilde{\nu} \leqq \tilde{\nu}_{max}$ in Eqs. (5.29) and (5.31).

The basic question for all practical considerations is that of the relation between $N_s$ and $N_{\tilde{\nu}}$. If we know this relation, we can predict $N_{\tilde{\nu}}$ from the experimental quantity $N_s$. In order to derive such a relation, we use Parseval's theorem [15] in the theory of Fourier transforms; this reads in our context

$$\int_{-s_{max}}^{+s_{max}} N^2(s)\, S^2(s)\, ds = 2 \int_0^{\tilde{\nu}_{max}} N^2(\tilde{\nu})\, d\tilde{\nu}\,. \tag{5.31}$$

With the help of Eq. (5.31), the required relation is easily established

$$2 s_{max} N_s^2 = 2 \tilde{\nu}_{max} N_{\tilde{\nu}}^2 \tag{5.32}$$

In practice, the RMS value $N_s$ of noise in the interferogram has to be considered a property of the detector and the electronic system. And generally the assumption holds that the RMS-value of the noise produced by such a system is proportional to the square root of the bandwidth of the system

$$N_s \sim \sqrt{\Delta f} \quad \text{or} \quad N_s^2 \sim \Delta f \tag{5.33}$$

That means that the noise $N(s)$ in the interferogram can be reduced by reducing $\Delta f$. But if we do this, we have to reduce also the speed $v$ and to increase the time $T$ for scanning the interferogram. The maximum wave number $\tilde{\nu}_{max}$ is usually a fixed quantity depending only on the spectroscopic problem under investigation. And the relation between $f_{max}$ and $\Delta f$ must not be changed according to our above arguments. Therefore, a reduction of $\Delta f$ requires the same reduction of $f_{max}$ and, consequently, that of $v$ ($f_{max} = 2 v \tilde{\nu}_{max}$!). Otherwise, the signatures of radiation with wavenumbers close to $\tilde{\nu}_{max}$ would be attenuated to an untolerable extent in recording the interferogram with the system of reduced bandwidth $\Delta f$. Now, we can relate the constant speed $v$ of the movable mirror to the maximum path difference $s_{max}$ and the scanning time $T$. It is

$$s_{max} = 2\, vT \quad \text{or} \quad v = \frac{s_{max}}{2T}$$

in the case of a single-sided interferogram. Combining this result with Eqs. (5.30) and (5.33), we obtain $\Delta f \sim \frac{s_{max} \cdot \tilde{\nu}_{max}}{T}$ and

$$N_s \sim \sqrt{\frac{s_{max} \cdot \tilde{\nu}_{max}}{T}}. \tag{5.34}$$

With Eq. (5.34), a connection is established between the parameters $s_{max}$, $\tilde{\nu}_{max}$, and $T$ of an actual experiment and the RMS-value of the noise in the interferogram. According to Eq. (5.32), the corresponding RMS-value $N_{\tilde{\nu}}$ of the noise in the spectrum is

$$N_{\tilde{\nu}} \sim \sqrt{\frac{s_{max}^2}{T}}. \tag{5.35}$$

Eqs. (5.34) and (5.35) show that we have to pay for better resolution with a decreased signal-to-noise ratio in Fourier spectroscopy, too. Eq. (5.34) tells us that $N_s$, the noise in the interferogram, is kept constant if with the resolution ($s_{max}$) the scanning time $T$ is linearly increased, *i.e.* the scanning speed is kept constant. But for $N_{\tilde{\nu}}$, the noise in the spectrum, another factor $s_{max}$ enters Eq. (5.35), originating essentially from the Fourier transform. This means that $N_{\tilde{\nu}}$

increases in proportion to $\sqrt{s_{\max}}$ for constant speed $v$ when the resolution is increased. The signal is not affected by increased resolution in Fourier transform spectroscopy. Therefore, we can write for the signal-to-noise ratio of the Michelson interferometer:

$$(S/N)_{\mathrm{M}} = \frac{S}{N_{\tilde{\nu}}} \sim \frac{\sqrt{T}}{s_{\max}}. \tag{5.36}$$

If we want to keep the signal-to-noise ratio constant when we increase the resolution by a factor of two, we have to increase $T$ by a factor of 4.

For comparison, in grating spectroscopy the signal is decreased with increased resolution. In order to double the resolution, we have to reduce the slit width by a factor of two. Then the signal is decreased by a factor of 4 according to Eq. (5.12). For the grating spectrometer also, the noise is assumed to be produced by the detector and the electronics; it is not affected by the decrease in the signal. And again we assume that the RMS-value of the noise $N$ (in the recorded spectrum!) is proportional to the bandwidth $\Delta f$ of the electronics. For Fourier spectroscopy, a suitable choice of the time constant $\tau$ is one third of time necessary to scan one sampling interval $\Delta s$ (cf. Section 4.6). For grating spectroscopy, the time constant $\tau$ should be equal to one third of the scanning time $t$ for one resolution width $\Delta\tilde{\nu}$. If the spectral range $\tilde{\nu}_{\min} \leqq \tilde{\nu} \leqq \tilde{\nu}_{\max}$ is scanned with constant speed in a total time $T$, we have

$$t/_{\Delta\tilde{\nu}} = T/_{(\tilde{\nu}_{\max}-\tilde{\nu}_{\min})} \,.$$

And with $\tau = \frac{1}{3}\, t$ and $\Delta f \sim \frac{1}{\tau}$ we obtain

$$\Delta f \sim \frac{1}{t} = \frac{\tilde{\nu}_{\max} - \tilde{\nu}_{\min}}{T \cdot \Delta\tilde{\nu}} \,.$$

The spectral range $(\tilde{\nu}_{\max} - \tilde{\nu}_{\min})$ is a fixed quantity for a given spectroscopic investigation. Therefore, we can drop this factor. The resolution width $\Delta\tilde{\nu}$ is proportional to the slit width $w$ [cf. Eq. (5.8)]. Thus our final result is

$$\Delta f \sim \frac{1}{T \cdot w} \,. \tag{5.36}$$

Eq. (5.36) means that the product $T \cdot w$ must not be changed without changing $\Delta f$. When the slit width $w$ is decreased and the resolution increased, finer details in the spectrum have to be transmitted by the electronics. If the scanning speed is not slowed down and the time $T$ enlarged, the finer details mean more rapid variations in the signal. These can only be transmitted when the bandwidth $\Delta f$ is increased. In other words, the finer details of the spectrum have to be resolved with respect to time, and the bandwidth $\Delta f$ or the scanning $T$ has to be increased. On these grounds, we can write for the RMS-value of the noise

$$N \sim \sqrt{\Delta f} \sim \frac{1}{\sqrt{T \cdot w}} \,. \tag{5.37}$$

As already mentioned, the signal S is proportional to $w^2$ [cf. Eq. (5.12)]. Therefore the signal-to-noise ratio for the grating spectrometer depends on w and T in the following way:

$$(S/N)_G \sim w^2 \sqrt{wT} = \sqrt{w^5 T} \,. \tag{5.38}$$

To keep this ratio constant for a resolution increased by a factor of two, we have to increase the scanning time $T$ by a factor of 32. This comparison again demonstrates the superiority of Fourier transform spectroscopy.

When an interferogram is sampled at equal increments $\Delta s$ of path difference, the accuracy of the procedure depends on the laser or Moiré system for measuring the path difference. And in reality, the sample values of the interferogram are not taken exactly at $s = n\Delta s$ but at the positions

$$s = n\Delta s + \varepsilon_n \tag{5.39}$$

where $\varepsilon_n$ is the statistical error in the determination of path difference. From this error, a random phase modulation originates in the interferogram

$$\bar{I}(n\Delta s + \varepsilon_n) = 2 \int_0^\infty I(\tilde{\nu}) \cos(2\pi\tilde{\nu} n\Delta s + 2\pi\tilde{\nu}\varepsilon_n)\, d\tilde{\nu} \,. \tag{5.40}$$

Since $\varepsilon_n$ is a small quantity, the regular and the random part of the interferogram Eq. (5.40) can be separated

$$\bar{I}(n\Delta s + \varepsilon_n) = \bar{I}(n\Delta s) + \frac{dI(n\Delta s)}{ds} \cdot \varepsilon_n \,. \tag{5.41}$$

In contrast to the detector noise $N(s)$ as discussed before, the sampling errors $\varepsilon_n$ introduce a signal dependent noise in the interferogram. The average of these errors is zero ($\overline{\varepsilon_n} = 0$). They can be assumed to be independent of each other and not correlated:

$$\overline{\varepsilon_m \cdot \varepsilon_n} = \begin{cases} \sigma^2 & \textit{if } m = n \quad \text{(RMS-value)} \\ 0 & \textit{if } m \neq n \quad \text{(no correlation)} \end{cases}$$

Here again, the RMS-value $\sigma$ is the interesting quantity. Bell and Sanderson [96] have shown the RMS-value of the corresponding noise in the spectrum to be

$$N_{\tilde{\nu}} = \sigma \sqrt{\frac{1}{2\tilde{\nu}_{max}} \int_{-\tilde{\nu}_{max}}^{+\tilde{\nu}_{max}} (2\pi\tilde{\nu})^2 [I(\tilde{\nu})]^2 \, d\tilde{\nu}} \,. \tag{5.42}$$

Eq. (5.42) demonstrates that, in this case, $N_{\tilde{\nu}}$ is not independent of the signal but proportional to a quantity which can be called the RMS-value of $\tilde{\nu} \cdot I(\tilde{\nu})$ of the spectrum under investigation. At last, the question arises what the experimenter can do to keep the noise in the recovered spectrum with RMS-value $N\tilde{\nu}$ as small as possible. For statistical errors of course, no error correction is possible as in the case of phase errors. But from our considerations, we may extract some guide lines along which one can try to minimize the influence of noise

a) The detector noise is proportional to $s_{\max}$. Therefore, it is not advisable to increase $s_{\max}$ and the resolution to an unnecessary extent. If once all details of the spectrum have been resolved any further increase of the resolution will not improve the spectrum but increase the noise.

b) With respect to sample errors, the user of a commercial instrument depends on the system for measuring the path difference. And with a He-Ne-laser controlling the path difference, a frequency precision better than 0.01 $cm^{-1}$ is achieved according to the data provided by the manufacturer. Under these circumstances, the statistical errors in sampling the interferogram and their RMS value $\sigma$ are probably too small to contribute considerably to the nosie in the spectrum, *i.e.* to $N_{\tilde{\nu}}$.

c) In the very far-infrared, the experimenter should be aware of the possibility that source fluctuations may contribute to the noise at rather slow scanning speeds. If this is the case or if this is suspected, it is advisable to scan the interferogram with a higher speed, to repeat this several times and to average the data.

## 6. Commercial Instruments

### 6.1 Survey of the Instruments

This Section compiles information on a number of commercially available Fourier transform spectrometers. Though this compilation is not complete, the interested reader will find here descriptions and technical data of altogether thirteen instruments from six manufacturers. Especially those instruments have been included here which were recently developed or improved. Therefore, this introduction to commercial instruments in the field of Fourier transfrom spectroscopy is believed to be a useful completion of a similar one published about 4 years ago in Ref. [4].

Before entering into the details of instrumentation the Fourier spectrometers are listed together with the manufacturers and their addresses. In some cases, also the company is given which sells the instruments in the Federal Republic of Germany:

1) Far Infrared Spectrometer System Model IR-720
Manufacturer: Beckman-RIIC Ltd.
Eastfield Industrial Estate, Glenrothes KY7 4NG, Scotland
Address in the Federal Republic of Germany:
Beckman Instruments, Technisches Büro München
D-8000 München 60, Otto-Engl-Platz 5

2a) Far Infrared Fourier spectrometer IFS 114

2b) Infrared Fourier spectrometer IFS 115
Manufacturer: Bruker-Physik AG
D 7501 Karlsruhe-Forchheim, Silberstreifen

3) FS 4000 Far Infrared Fourier Transform Spectrometer
Manufacturer: Coderg
15 Impasse Barbier, F-92110 Clichy
Authorised dealer in the Federal Republic of Germany:
Amko GmbH & Co, KG,
D-2082 Tornesch, Lindenweg 53

4) Digilab Fourier Spectrometers
(FTS 10, FTS 14, FTS 15, FTS 20, FTS 16)
Manufacturer: Digilab Inc.
237 Putnam Avenue, Cambridge/Mass 02139, USA
Authorised dealer in the Federal Republic of Germany:
Cambridge Instrument Company GmbH,
D-4600 Dortmund, Harnackstraße 35—43

5a) Fourier Spectrometer MIR 20

5b) Fourier Spectrometer MIR 160

5c) Far Infrared Fourier Spectrometer FIR 30
Manufacturer: Polytec GmbH
D-7517 Waldbronn-Karlsruhe (Reichenbach), Siemens-Straße

6) Model ST-10 GC-IR automated System
Manufacturer: Spectrotherm Corporation
3040 Olcott Street, Santa Clara, California 95051, USA
Authorised dealer in the Federal Republic of Germany:
Amko GmbH & Co, KG,
D-2082 Tornesch, Lindenweg 53

The basic properties of these commercial instruments and their prices have been collected in Table 2. From the spectral range covered in the standard version (without extensions), we may distinguish two categories of instruments:
a) instruments for the far-infrared (No. 1, 2a, 3, 4e, 5c)
b) instruments for the middle- and near-infrared (No. 2b, 4a—4d, 5a, 5b, 6).

Among the five far-infrared Fourier spectrometers, three are slow-scan instruments with a Golay detector and a Moiré-system controlling the path difference. Their design is more or less taylormade for the far-infrared where the radiation energy is small and where relatively large scanning times are needed to obtain an agreeable signal-to-noise ratio. In this range, we have relatively large sampling intervals $\Delta s$ (cf. Table 1 in Section 4.6), and the accuracy of a Moiré-system is sufficient in measuring path differences. The other two far-infrared instruments are rapid-scan instruments. Their design is essentially the same as that of the instruments in the near- and middle-infrared, except that the source is a mercury arc and not a glower. They employ a fast pyroelectric detector and a He-Ne-laser as the reference system. And it is well known that this basic

Table 2. Survey of commercial Fourier transform spectrometers

| No. | Instrument | Spectral range[1] | Source | Detector | Reference system[2] | Maximum resolution[3] | Price[4] |
|---|---|---|---|---|---|---|---|
| 1 | Beckman IR 720 | 500—10 $cm^{-1}$<br>(600—10 $cm^{-1}$) | Mercury arc | Golay | Moiré | $\Delta\tilde{\nu} = 1.56\ cm^{-1}$ | DM 122070.— |
| 2a | Bruker IFS 114 | 1000—10 $cm^{-1}$<br>(4000—10 $cm^{-1}$) | Mercury arc | Pyroelectric | He-Ne-Laser | $\Delta\tilde{\nu} = 0.5\ cm^{-1}$ | Prices are of the same order of magnitude as those for comparable instruments. |
| 2b | Bruker IFS 115 | 4000—400 $cm^{-1}$<br>(20000—10 $cm^{-1}$) | Glower | Pyroelectric | He-Ne-Laser | $\Delta\tilde{\nu} = 2\ cm^{-1}$ | |
| 3 | Coderg FS 4000 | 750—15 $cm^{-1}$ | Mercury arc | Golay | Moiré | $\Delta\tilde{\nu} = 0.5\ cm^{-1}$ | ca. DM 190000.— |
| 4a | Digilab FTS 10 | 4000—400 $cm^{-1}$ | Glower | Pyroelectric | He-Ne-Laser | $\Delta\tilde{\nu} = 2\ cm^{-1}$ | DM 150000.— |
| 4b | Digilab FTS 14 | | | | | 0.4 $cm^{-1}$ | DM 270000.— |
| 4c | Digilab FTS 15 | 4000—400 $cm^{-1}$<br>(10000—10 $cm^{-1}$) | Glower | Pyroelectric | He-Ne-Laser | 0.25 $cm^{-1}$ | DM 289000.— |
| 4d | Digilab FTS 20 | | | | | 0.125 $cm^{-1}$ | DM 310000.— |
| 4e | Digilab FTS 16 | 450—10 $cm^{-1}$ | Mercury arc | Pyroelectric | He-Ne-Laser | 2 $cm^{-1}$ | DM 198000.— |
| 5a | Polytec MIR 20 | 4000—400 $cm^{-1}$<br>(10000—10 $cm^{-1}$) | Globar | Pyroelectric | He-Ne-Laser | $\Delta\tilde{\nu} = 0.5\ cm^{-1}$ | DM 252000.— |
| 5b | Polytec MIR 160 | | | | | 0.06 $cm^{-1}$ | DM 285000.— |
| 5c | Polytec FIR 30 | 1000—10 $cm^{-1}$ | Mercury arc | Golay | Moiré | 0.6 $cm^{-1}$ | DM 177000.— |
| 6 | Spectrotherm ST 10 | 4000—700 $cm^{-1}$ | Glower | (Hg, Cd) Te[5] | He-Ne-Laser | $\Delta\tilde{\nu} = 2\ cm^{-1}$ | ca. DM 100000.— |

[1]) Standard version, in brackets: optional.
[2]) For measuring the path difference.
[3]) As achievable with standard version.
[4]) Costs of complete standard spectrometer with computer.
[5]) Cooled by liquid nitrogen.

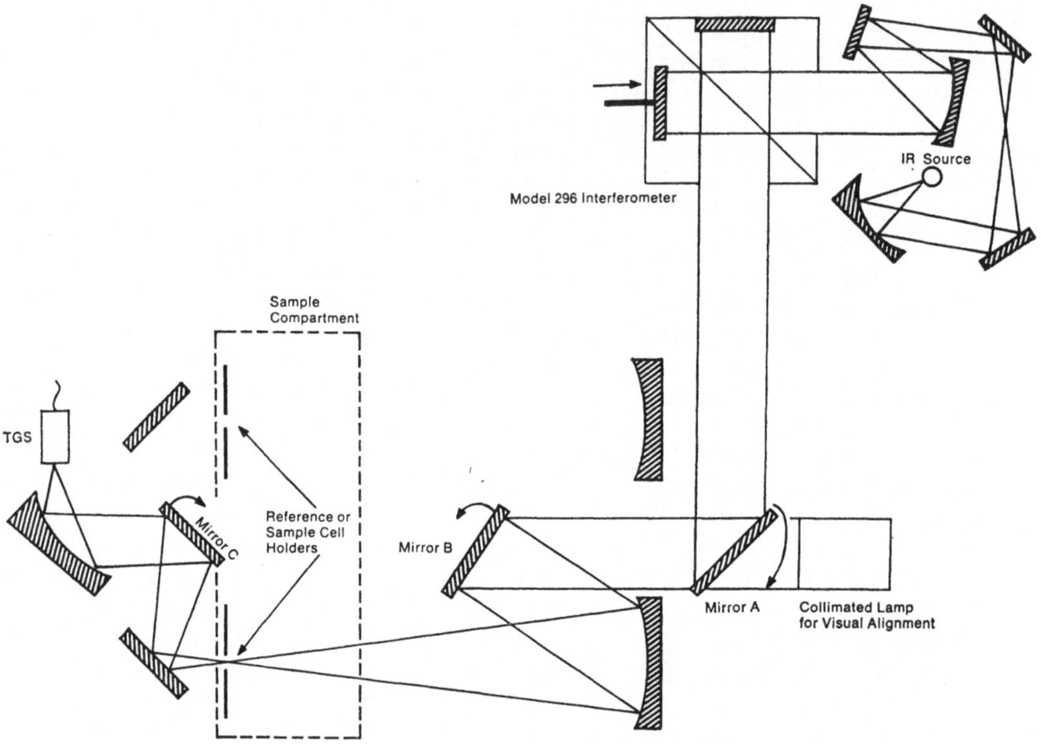


Fig. 44. Optical diagram of the Digilab FTS 14 Fourier spectrometer (No. 4b in Tables 2, 3, 4)

concept for the near- and middle-infrared instruments has been proved a success. And especially for the "fingerprint" region of the chemists (400—4000 $cm^{-1}$), a great variety of instruments is available on the market. Their prices range from about DM 100,000.— to DM 300,000.—. The factor of three between the cheapest and the most expensive instrument is mainly due to the different maximum resolution achievable with the instrument and to the different equipment with respect to computer memory etc. Most of these instruments may be used in connection with gas chromatography. The instrument No. 6 (Spectrotherm ST 10) is especially designed for this purpose. The price differences between the far-infrared instruments can also be related to different maximum resolution, computer outfit etc.

## 6.2 The Optical Layout

The optical layouts of the instruments are very similar, especially those of the instruments for the middle- and near-infrared. For comparison, the optical diagrams of three such instruments are reproduced in Figs. 44 (No. 4b, Digilab FTS 14), 45 (No. 2b, Bruker IFS 115), and 46 (No. 5b, Polytec MIR 160). Clearly, the essential part of the optical layout is the Michelson interferometer with the

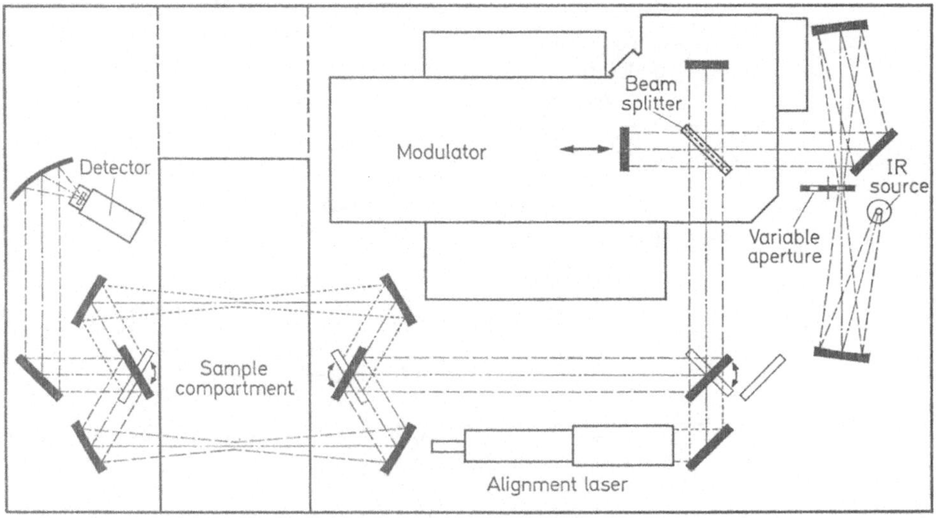


Fig. 45. Optical diagram of the Bruker IFS 115 Fourier spectrometer (No. 2b in Tables 2, 3, 4)

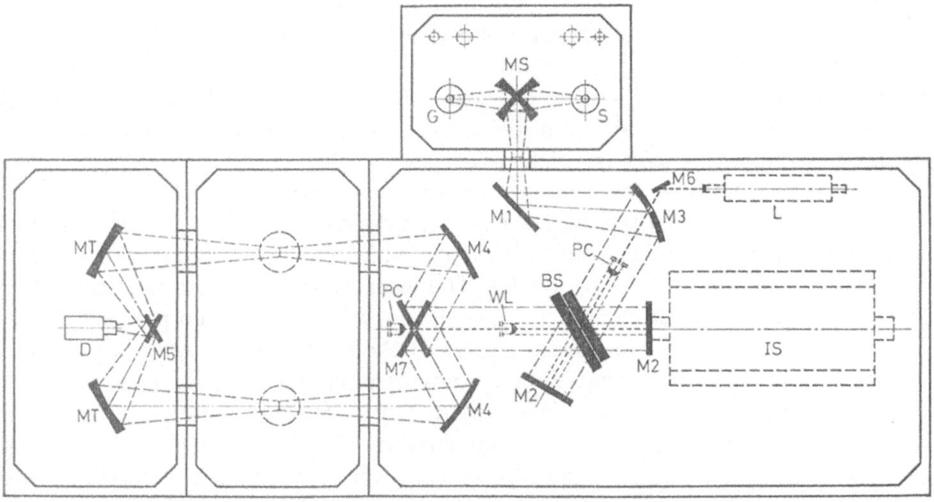


Fig. 46. Optical diagram of the Polytec MIR 160 Fourier spectrometer (No. 5b in Tables 2, 3, 4). M 1, M 2, M 5, M 6, M 7: plane mirrors; M 3, M 4: paraboloid mirrors; MS: spherical mirror; MT: toroid mirrors; G: Globar source; S: high pressure Hg-lamp; L: He-Ne-laser; IS: Interferometer scanner; BS: beampslitter; PC: photo-cell; D: pyroelectric detector; WL: white light source

beamsplitter, with the movable and the fixed mirror. In the spectral range above 500—1000 $cm^{-1}$, the beam splitters are dielectric films on a susbstrate (cf. the principal considerations in Section 4.1). The materials employed as beamsplitters in commercial instruments are compiled in Fig. 47. It should be noted that for each

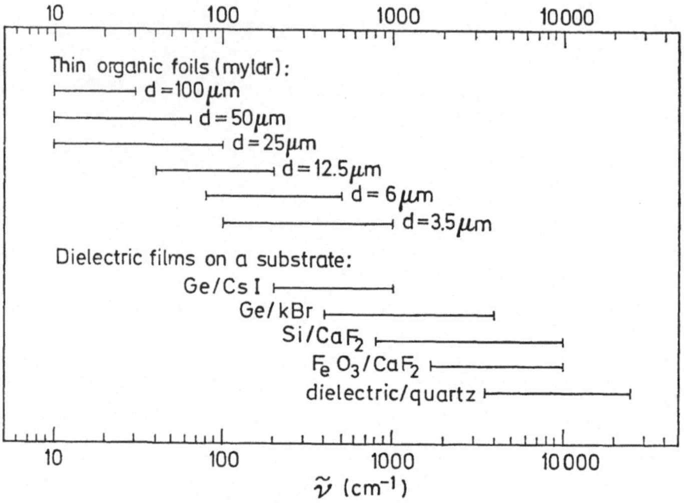


Fig. 47. Several beamsplitters for Fourier spectrometers and the spectral range for which they are used predominantly, especially in commercial instruments

beamsplitter the spectral range is indicated in the figure in which it is used in the instruments and in which range it probably is the most suitable. But that does not necessarily mean that the beam splitter efficiency is more or less zero outside the quoted range. In Figs. 45 and 46 the plate is clearly indicated that compensates the optical effects of the substrate of the beam splitter. The effective thin film is between the two relatively thick plates.

As a special feature, the optical diagram of the Bruker IFS 115 (cf. Fig. 45) shows a variable source diaphragm at an extra focus. The instrument is equipped with a helpful alignment laser the radiation of which can be used to align the interferometer as well as the sample optics. The laser radiation is sent into the interferometer or into the sample optics or is kept outside the radiation path by simply turning a plane mirror. The spectrometers of the Digilab FTS-series have a conventional collimated light source for alignment purposes (cf. Fig. 44).

A common property of most instruments is that they are built up by modular units: source and interferometer compartment, sample chamber and detector housing (cf. Figs. 45, 46). In the sample compartment, there are generally two foci. One of the foci is used for the sample spectrum, and the other for the reference or background spectrum. Either the sample focus or the other one is incorporated in the radiation path by turning two plane mirrors (cf. Figs. 44—46). This operation is usually automatized and controlled by the computer system of the spectrometer. The size of the sample focus varies from 3 mm to 13 mm diameter, somewhat differing between different instruments and also depending on the diameter of the diaphragm used at the source (cf. Table 3). The sample chamber of all instruments is large enough to position there a variety of accessories as cryostats, gas cells, superconducting magnets, etc. In addition to transmission measurements, also reflection measurements (by using a reflection attachment) and emission measurements can be performed. Such a reflection attachment is shown in Fig. 50 with two spherical and four plane mirrors providing a focus for the reflection sample and two other foci, at one of which a diaphragm may be placed to limit the size of the radiation spot on the sample.

The optical diagram of the Polytec MIR 160 (cf. Fig. 46) shows very instructively the He-Ne-laser reference system. The laser radiation enters through a hole in the center of mirror M3 the Michelson interferometer. The interference fringes monitoring the travel path of the movable mirror M2 are recorded by the photo cell PC (close to turnable mirror M7). As we have seen in Section 2.1 at the beginning of our considerations, the interferogram of monochromatic radiation is a $\cos^2$-curve (see Fig. 2). It has no particular signature at $s=0$ equivalent to the the grand maximum of an interferogram corresponding to a broad continuous spectrum (cf. Fig. 10 in Section 3.2). In order to provide the information about zero path difference, most instruments use a white light reference source (WL in Fig. 46) in addition to the He-Ne-laser. Its interferogram is recorded by a second photo cell (between BS and M3 in Fig. 46). The grand maximum of this interferogram indicates the position $s=0$ the knowledge of which is essential for a correct evaluation of the Fourier transformed spectrum as it was pointed out in connection with phase errors in Section 5.3.

A very peculiar optical layout is found in the Spectrotherm ST 10 interferometer (see Fig. 48). The radiation emittted by the source is collimated by mirror T1 and reflected by mirror M1. At the beam splitter, part of the radiation is reflected back to the other half of M1 and the other part transmitted to the movable mirror M2. Both parts are reflected by mirrors M1 add M2, resp., to the beamsplitter which again partly transmits and partly reflects the two beams. From each beam, nearly one half is refocussed by mirror T2 into the reference cell and the other half by mirror T2′ into the sample cell. In this way, radiation is passed simultaneously through the reference cell and through the sample cell. In both beams, the radiation shows interference fringes according to the path difference, and an interferogram can be recorded. The interferogram of the radiation through the sample cell is that usually obtained while the interferogram of the radiation through the reference cell is the complementary one which is usually reflected back to the source [see Eq. (4.1) in Section 4.1]. Now, the radiation through the sample cell and through the reference cell are both focussed on the same detector. In Section 4.1 was pointed out that the mean level drops out when the difference is formed of the interferogram transmitted by the Michelson [Eq. (3.3)] and of the reflected [cf. Eq. (4.1)]. Now, when the sum is formed of these two by using one detector, the oscillatory parts of the interferograms drop out. The sum of the mean levels is a constant signal independent of path difference to which the detector does not respond. But if there is absorption in one of the cells, the two oscillatory parts of the interferograms do no longer completely compensate, and, as a result of this, the detector records an interferogram signal. In this way, the Spectrotherm ST 10 can be called a "real" double-beam Fourier spectrometer (cf. Ref. [43]) though some mathematical corrections may be necessary (cf. Ref. [99]). At last it remains to be mentioned that a great number of the commercial instruments for the middle- and near-infrared can be flushed with nitrogen or dry air to prevent atmospheric absorption and that only some of them can be evacuated.

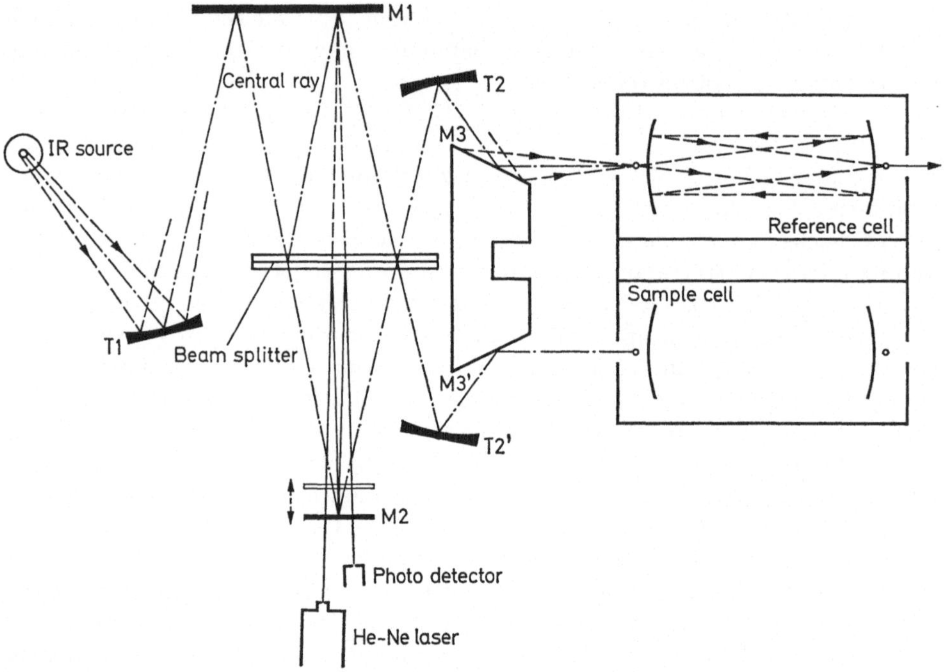


Fig. 48. Optical diagram of the Spectrotherm SP 10 Fourier spectrometer (No. 6 in Tables 2, 3, 4). M 1, M 3, M 3′: plane mirrors; T 1, T 2, T 2′: toroid mirrors; M 2: moving mirror

Most of the preceding considerations with respect to the optical layout of Fourier spectrometers apply equally well to the far-infrared instruments. One essential difference is that in this range self-supporting thin films are used as beamsplitters (cf. discussion in Section 4.1 and see Fig. 47). As there is an extremely strong water vapor absorption in the 100 $cm^{-1}$ region, most of the far-infrared instruments are vacuum instruments and can be evacuated to remove even small traces of water vapor. And the thin film beamsplitters are suitable only for a limited range (cf. Fig. 47) and have to be changed frequently. In this situation, it is helpful for the experimenter to have various beam splitters mounted on a wheel which can be rotated by means of a remote control thus changing beamsplitters automatically, without breaking the vacuum. Such a rotating beamsplitter wheel is provided in the Coderg FS 4000 and in the Polytec FIR 30 (cf. optical diagram in Fig. 49) Fourier spectrometers. Fig. 49 shows also the chopper which is used for modulating the radiation at a certain frequency in all slow-scan instruments and which is not found in the rapid-scan instruments. In the sample chamber of the Polytec FIR 30, there is only one focus. Now, the reflection attachment is included regularly in this spectrometer. And by means of the mirrors M5 (cf. Fig. 49), the optical arrangement in the sample chamber can be used for reflection and transmission measurements in the same way as if there were two foci. This is indicated in the upper part of Fig. 49. The two mirrors M5 can be turned automatically by operating a switch on the control board.

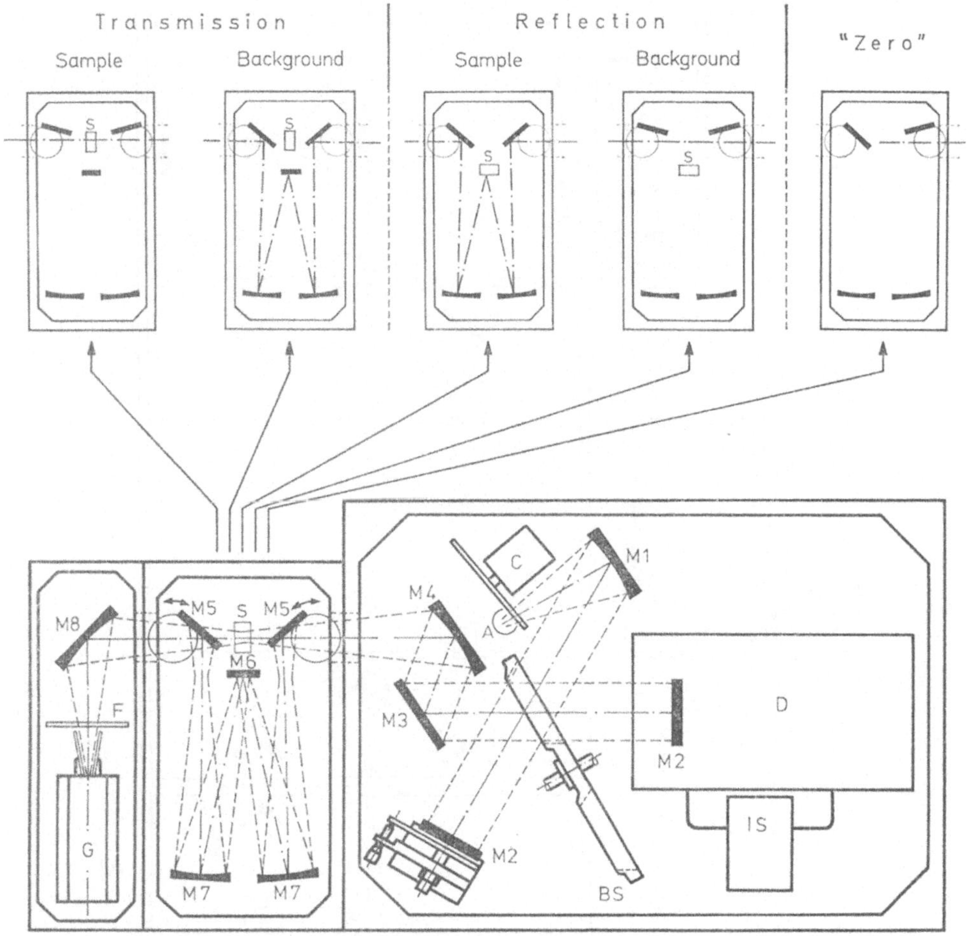


Fig. 49. The Polytec FIR 30 Fourier spectrometer (No. 5c in Tables 2, 3, 4) Optical diagram (lower part) and the possibilities of using the sample chamber (upper half). M 2, M 3, M 5, M 6: plane mirrors; M 1, M 4: paraboloid mirrors; M 7; spherical mirror; M 8: elliptical mirror; C: Chopper; S: high pressure Hg-lamp; BS: beamsplitter; D: mirror drive; IS: Moiré system; G: Golay detector; S: sample

Another example of an far-infrared Fourier spectrometer is the Bruker IFS 114 (cf. Fig. 50). In contrast to the Polytec FIR 30, the Coderg FS 4000 and the Beckman FS 720, this spectrometer is a rapid-scan instrument. But the consequences of rapid and slow scanning will be discussed later. Here we shall concentrate on the optical layout. From Figs. 49 and 50, it is evident, that the interferometers are constructed as a modular system and that, by inserting some windows, the compartments of the several modules can be evacuated separately. This is of particular interest for the sample chamber. There, it is an advantage to be able to exchange samples without breaking the vacuum of the whole instrument and to have to reevacuate only the sample chamber. The different modules of the

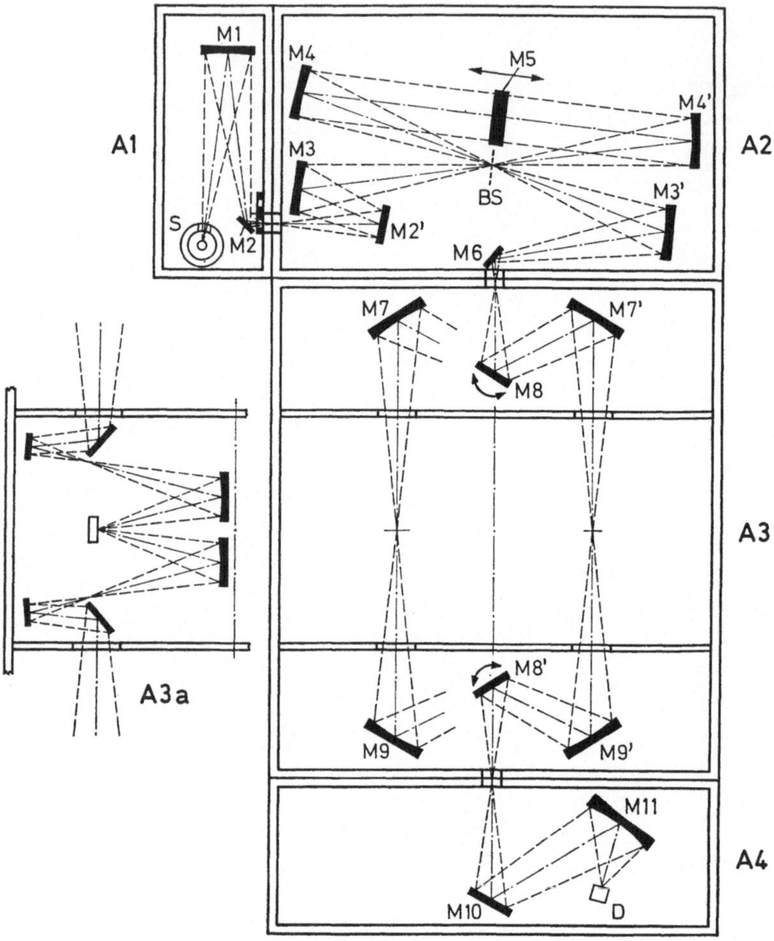


Fig. 50. Optical diagram of the Bruker IFS 114 Fourier spectrometer (No. 2a in Tables 2, 3, 4), M 1, M 3, M 4, M 7, M 9, M 11: concave mirrors; M 2, M 5, M 6, M 8, M 10: plane mirrors; S: Source; BS: beamsplitter; D: detectors

Bruker IFS 114 are marked in Fig. 50. For the sample compartment (A 3), the reflection attachment (A 3a) is shown in Fig. 50 which can be inserted at one of the two foci. Characteristic and outstanding is the design of the Michelson interferometer in this instrument (A 2 in Fig. 50). The radiation emitted by the source is not collimated at first but focussed on the beam-splitter. The angle of incidence on the beam-splitter is rather small (nearly normal incidence). This is advanttageous with respect to the polarization properties of the beam-splitter. We recall from Section 4.1 and Fig. 15 that the efficiency (4 RT) of thin film beamsplitters depends on its thickness, on its refractive index, on the wave number of the radiation, and, last but not least, on the polarization of the light. The latter dependence is rather drastic for angles of incidence near 45° which are close to the Brewster angle for many organic films. In Section 4.1, it was pointed out that the intensity of radiation polarized perpendicular to the plane of incidence is about seven times larger than that of radiation polarized parallel to the plane of incidence (cf. Fig. 15, angle of incidence 45° and $n = 1.5$). For nearly normal incidence, the intensity is equal for the two directions of polarization. Thus, the recorded interferogram does not depend on the polarization in this case. Moreover, metal screen beam-splitters (cf. Fig. 16) which can be used successful in the very far-infrared have rather complex polarization properties except they are used at nearly normal incidence [100]. And beam-splitters with comparatively small areas can be employed in this instrument since they are placed at a focus. The radiation reflected and transmitted by the beam divider (BS in Fig. 50) is then collimated by the mirrors M 4 and M 4′, respectively, and sent to the movable mirror M 5 which has two reflecting, parallel surfaces. Each of the partial beams is reflected back to the collimator mirrors (M 4 and M 4′) and is focussed on the beam-splitter (BS). There, they are partly transmitted and reflected as usual. In this interferometer, the optical path difference between the two partial beams is fourtimes the distance travelled by mirror M 5 away from the position $s = 0$ (white light position). In usual Michelson interferometers, the path difference $s$ is only twice the path travelled by the movable mirror as was pointed out in Section 4.2. Here, the factor 4 can be understood as follows: If mirror M 5 is moved a distance $\Delta x$, the optical path of the one partial beam is increased by $2\,\Delta x$ while that of the beam reflected at the other side of M 5 is decreased by $2\,\Delta x$. And the resulting change in path difference between the two partial beams is $s = 4 \cdot \Delta x$. Finally, it seems worth noticing that there is sufficient space in the detector compartment (A 4 in Fig. 50) for the pyroelectric detector and for a second detector, *e.g.* a cooled one for the very far-infrared.

### 6.3 Interferometer Data

A number of data has been collected in Table 3 which are to be discussed in this section. These data concern the optics, the aperture and the throughput of the interferometer in various Fourier spectrometers. In addition, it seems to be of interest to comment on the mode of propulsion, the speed and the maximum travel path of the movable mirror.

For all Fourier spectrometers discussed here, the aperture or the $f$-number is listed in Table 3. As usual in optical instruments, the $f$-number is the ratio of the diameter of a lens or a concave mirror and its focal length. For example, $f/4$ means that the focal length of the lens or mirror is larger by a factor of 4 than the diameter. In case of a Michelson, interferometer, let us refer the $f$-number to the collimator mirror, *i.e.* its diameter and its focal length. Obviously, the ratio of these two quantities is a measure of the solid angle $\Omega_c$ subtended by the collimator mirror as was pointed out in Section 5.1 in context with the throughput of the interferometer. We recall that

$$\Omega_c = \frac{A_c}{f^2} = \frac{\pi R^2}{f^2} \tag{6.1}$$

Table 3. Some data of the interferometers in commercial instruments

| No. | Instrument | Diameter of diaphragm at source (cm) | Aperture of optics[1] | Diameter of collimator mirrors (cm) | Throughput (Sterad-$mm^2$)[2] | Frequency precision ($cm^{-1}$) | Propulsion (of the movable mirror) | Speed (of the movable mirror) (mm/sec) | Maximum path (of the movable mirror) (cm) | Maximum resolution[2] ($cm^{-1}$) |
|---|---|---|---|---|---|---|---|---|---|---|
| 1 | Beckman IR-720 | 0.3—1.0 | $f/2$ | — | 1.4—15 | | Synchronous motor | $(0.5—50) \cdot 10^{-3}$ | 0.5 | 0.1 |
| 2a | Bruker IFS 114 | 1.0 | $f/3.7$ | 5 | 5 | <0.1 | Linear motor (air bearing) | 0.2—15 | 2.5[3] | 0.1 |
| 2b | Bruker IFS 115 | 0.2—1.0 | $f/3.7$ | 5[4] | 0.1—5 | <0.01 | | 0.2—25 | 8.3 | 0.06 |
| 3 | Coderg FS 4000 | 0.3—1.0 | $f/2$ | — | 1.4—15 | | Synchronous motor | $(1—50) \cdot 10^{-3}$ | 5 | 0.1 |
| 4a | Digilab FTS 10 | | | | | | | | 0.25 | 2.0 |
| 4b | Digilab FTS 14 | | | | | | | | 1.25 | 0.4 |
| 4c | Digilab FTS 15 | 0.8 | ca. $f/5$ | 5 | 1.5 | <0.01 | Linear motor (air bearing) | 1.56 | 2.0 | 0.25 |
| 4d | Digilab FTS 20 | | | | | | | | 4.0 | 0.125 |
| 4e | Digilab FTS 16 | | | | | | | | 2.0 | 0.25 |
| 5a | Polytec MIR 20 | 0.3—1.0 | $f/4$ | 5 | 0.3—4 | <0.01 | Linear motor | 2.5; 25 | 1.0 | 0.5 |
| 5b | Polytec MIR 160 | | | | | | | 2.5—40 | 8.0 | 0.06 |
| 5c | Polytec FIR 30 | 0.3—1.0 | $f/2$ | 6.5 | 1.4—15 | | Synchronous motor | $(1—25) \cdot 10^{-3}$ | 5.0 | 0.1 |
| 6 | Spectrotherm ST 10 | ≈0.1 | ca. $f/4$ | — | ≈0.04 | | "Porch-swing design" | 1.29 | 0.26 | 1.94 |

[1] $f/2$ means the diameter of the collimator mirror is half of its focal length.
[2] As evaluated from the maximum path of the movable mirror.
[3] $s_{max} = 4 \cdot 2.5\ cm^{-1}$ (special interferometer design, see text).
[4] Optional: $f/2$ optics with mirrors of 10 cm diameter.

where $R$ and $f$ are the radius and the focal length of the collimator mirror, respectively. Now, if we have an $f$-number $f/n$ for an interferometer the radius is $R = \frac{f}{2n}$, and we obtain for the solid angle

$$\Omega_c = \pi \left(\frac{f}{2n}\right)^2 \Big/ f^2 = \frac{\pi}{4n^2} \tag{6.2}$$

In the far-infrared region, energy is so badly needed that the slow-scan spectrometers especially designed for this range have a $f$-number $f/2$ (instruments No. 1, 3, and 5c in Table 3). The $f$-numbers of the other interferometers range from $f/3.7$ to $f/5$. These interferometers have been designed for the middle- und near-infrared (instruments No. 2b, 4a—d, 5a, 5b, and 6 in Table 3) or are rapid-scan spectrometers for the far-infrared (No. 2b and 4e in Table 3). The diameters of the collimator mirrors are usually about 5 cm. It should be noted that optionally a $f/2$ optics with collimator mirrors of 10 cm diameter can be provided for the Bruker IFS 115 Fourier spectrometer (No. 2b in Table 3). In order to evaluate the throughput of all these instruments, we need the characteristic area $A_s$ [cf. Eq. (5.2) in Section 5.1], in addition to the solid angle $\Omega_c$. In most interferometers, this characteristic area is that of the variable diaphragm between source and collimator mirror. Usually, its diameter ranges from 3 mm to 1 cm. From these data, the throughput $E_M = A_s \cdot \Omega_s$ can be evaluated. For the more luminous far-infrared spectrometers, the values of $E$ range from 1—15 sterad · mm², and for the other instruments, from .05—5 sterad · mm² (cf. Table 3). In this comparison, we realize that the values quoted to be typical in section 5.1 are typical values for the far-infrared ($f$-number $f/2$, source diameter 10 mm). And it should be noted that the values for the Fourier spectrometer and the grating spectrometer apply more to home-made instruments for the very far-infrared than to commercial instruments in the middle- and near-infrared. As already mentioned in other context, the optical path difference is controlled and measured by means of a He—Ne-laser in a number of instruments (No. 2a, b, 4a—e, 5a, b, and 6 in Table 3, cf. also Table 2). For these instruments, a frequency precision better than .01 $cm^{-1}$ can be guaranteed. This is due to the high stability of such a laser and also due to the fact that the laser frequency is independent of temperature and other influences. For a Moiré system on the other hand, the precision usually is given by that of a ruled glass rod which has a rather small but still finite thermal expansion. And the same argument holds for the grating in a grating spectrometer.

Now, let us concentrate on the mode of propulsion of the movable mirror in various Fourier spectrometers. Among the discussed 13 instruments, there are three spectrometers (for the far-infrared!) where the movable mirror is driven by a synchronous motor (No. 1, 3, and 5c in Table 3) and where the speed of the mirror can be varied from 1 to 25 $\mu$m/sec. These are the socalled "slow-scan" instruments. They employ a slow Golay-detector (cf. Table 2), and the radiation of the source is chopped at a certain frequency as was pointed out already in context with the optical diagram (cf. Fig. 49). In the other ten instruments listed in Table 3 (No. 2a, b, 4a—e, 5a, b, and 6), the movable mirror is driven by an electromagnetic device like a linear motor and where the speed can be varied from 1 to 25 mm/sec. In a rapid-scan instrument therefore, the movable mirror is faster by about a factor of thousand in comparison to slow-scan spectrometers. The friction problems arising for the high speeds are overcome by "air bearing" and similar constructions. With respect to the different speeds, we recall that radiation of wavenumber $\tilde{\nu}$ is modulated at a frequency [of Eq. (4.3) in Section 4.2].

$$f = 2v\tilde{\nu}$$

when the mirror is driven at constant speed $v$. At a speed of 1 $\mu$m/sec, the modulation frequencies are .004, .06 and .2 Hz for $\tilde{\nu} = 20$, 300, and 1000 cm$^{-1}$, respectively. At a speed of 1 mm/sec on the other hand, the modulation frequencies are 10, 160, and 800 Hz for $\tilde{\nu} = 50$, 800, and 4000 cm$^{-1}$, respectitely. From this comparison, it is obvious that we do not need a fast detector in a slow-scan instrument. The modulation frequencies from the interference fringes are almost d.c. electric signals. But for reasons of thermal drift and other stability problems, usual a.c. amplifiers are preferred to d.c. amplifiers for the amplication of small signals it is the case for far-infrared spectrometers. Therefore it is advisable to chop the radiation at a frequency which is sufficiently low for the Golay detector and not too low for an a.c. amplifier, *i.e.* a frequency between 10 and 15 Hz. A tuned amplifier can be used, and a bandwidth of about .5 Hz will be required to record an interferogram (with frequencies .004—.2 Hz) without distortion or attenuation of the high frequency components (cf. the considerations with respect to the electronic bandwidth in Sections 4.6 and 5.4). For a rapid-scan instrument on the other hand, a bandwidth of one or two kilohertz is necessary for the electronic system, and a fast pyroelectric detector has to be used (cf. Table 2). The modulation frequencies from the interferogram itself (10—800 Hz) are sufficiently high that usual a.c.-amplifiers can be used without employing a chopper. If the radiation were to be chopped in this case, a rather high frequency of about 10 kHz had to be applied where other problems could arise.

In this context, it is perhaps worth to mention that both, the Golay detector and the pyroelectric one, are thermal detectors. Both can be used without cooling at room temperature. The difference in speed between these two can be explained in terms of their effective time constants. In the Golay detector the radiation is absorbed in a thin film which is positioned inside a small gas chamber. Via the absorbing film, the radiation energy heats the gas, and the thermal expansion of the gas is finally transferred to an electrical signal. The time constant of the Golay detector is dependent on the thermal capacity and on the thermal conductivity of the system. A typical value of the time constant is $\tau = 15$ m sec [39]. The pyroelectric detector consists of a thin slice of a pyroelectric crystal like TGS (triglycine sulfate), SbSI or $BaTiO_3$. This slice is sandwiched between two electrodes one of them exposed to the radiation and absorbing the radiation if not a special coating is used for this purpose. These crystals exhibit a dielectric polarization which is temperature dependent, especially if the mean temperature is close to their ferroelectric Curie temperature. The absorption of radiation causes a temperature change, and this, via a change in the polarization, requires a change of the charges on the electrodes. The corresponding current is the electrical signal.

The time constant of this system depends on its electrical impedance. A typical value is $\tau = 40\ \mu$sec [39] for a TGS-detector which is smaller by a factor of about 400 than that of the Golay cell.

The last quantity to be discussed in this section is maximum path of the movable mirrors. In the slow-scan instruments, usually a lead-screw is employed to drive the mirror with the synchronous motor. And a maximum scan length of the movable mirror of 5—10 cm is achieved without any problems. In the case of rapid-scan instruments however, the customer has to pay for a larger maximum scan length. From a comparison of this quantity for various instruments in Table 3 (*e.g.* No. 4a—d and No. 5a, b) with the prices of the instruments in Table 2, we learn that with increasing prices also the maximum mirror path is increased which determines maximum resolution of the instruments as limited by the mechanics of the interferometer. We recall from our considerations in Sections 2.3, 3.2, 4.6 and 5.1 that, in Fourier spectroscopy, the resolution width is

$$\Delta \bar{\nu} = \frac{1}{s_{\max}} \tag{6.3}$$

[cf. Eq. (2.23) in Section 2.3] where $s_{\max}$ is the maximum optical path difference. By means of Eq. (6.3), the maximum resolution has been calculated and also listed in Table 3 for the various instruments. The least resolution is $\Delta \nu = 2.0\ \text{cm}^{-1}$ (instruments No. 4a and 6), the best value is $\Delta \nu = .06\ \text{cm}^{-1}$ (instruments No. 2b and 5b). From the viewpoint of practical applications, a resolution of about 1 $\text{cm}^{-1}$ is required for broad bands in solids (cf. Figs. 26 and 27) while a much better resolution is necessary for narrow lines as they happen to occur in gases for example (cf. Fig. 29).

### 6.4 Data about the Computers

This part of the comments on commercial instruments is devoted to the characteristic problem of Fourier spectroscopy, *i.e.* the necessity to subject the experimentally determined interferogram to the Fourier transform and to employ an electronic computer for this purpose. At first we shall concentrate on the more theoretical aspects of this problem. Among the slow-scan far-infrared instruments, two (No. 3 and 5c in Table 4) use the real time method (for explanation see Section 4.4) for evaluating the spectrum from the interferogram and one (No. 1 in Table 4) employs a wave analyzer, an analogue computer. But it should be noted, that the whole data processing system of the instrument No. 1 (Beckman IR 720) is a hybrid system. The interferogram is sampled, digitized and stored in a memory. From there, the data are transferred to the wave analyzer which operates in an analogue way. The method of Fourier transform in all the rapid-scan instruments (No. 2a, b, 4a—e, 5a, b, 6) is the fast Fourier transform (FFT or Cooley-Tukey algorithm). And it was pointed out in Sections 4.3 and 4.4 and in Appdx. 3 that this method is the most appropiate for these Fourier spectrometers. Except for the Beckman IR 720 with the wave-analyzer, for all instruments a mathematical phase error correction is included automatically in the Fourier transform program. As already mentioned in Section 4.4, the possibilities of a phase error correction are limited in the case of the real-time Fourier-transform.

Table 4. Fourier transform and computers in commercial instruments

| No | Instrument | Method of Fourier transform | Phase error correction | Computer memory[1] core | Computer memory[1] disk | Maximum number of data points[2] | Sampling interval $\Delta s$ ($\mu$m) | Maximum resolution[3] ($cm^{-1}$) |
|---|---|---|---|---|---|---|---|---|
| 1 | Beckman IR-720 | Wave anlyzer[4] | None | 1.6 K | — | $2 \times 0.8$ K[5] | 4, 8, 16, 32, 64 | 1.56 |
| 2a | Bruker IFS 114 | Cooley-Tukey (FFT) | Mathematical phase error correction | 16 K (80K) | (1.2M) | $2 \times 4$ K | $2^n \times 0.6328$ ($-1 \leqq n \leqq +4$) | 0.50 |
| 2b | Bruker IFS 115 | | | | | | | 1.98 |
| 3 | Coderg FS 4000 | Real time | Special electronic phase error correction | 4 K | — | $2 \times 1.5$ K | 5, 10, 20, 40, 80 | 0.45 |
| 4a | Digilab FTS 10 | Cooley-Tukey (FFT) | Mathematical phase error correction | 16K (32K) | — | $2 \times$ ca. 4 K | $2^n \times 0.6328$ | 1.98 |
| 4b | Digilab FTS 14 | | | 8K | 128K | $2 \times$ ca. 40 K | | (0.20)[6] |
| 4c | Dibilab FTS 15 | | | 8K | 1.2M | $2 \times$ ca. 500 K | | (0.02)[6] |
| 4d | Digilab FTS 20 | | | 8K | 1.2M | $2 \times$ ca. 500 K | | (0.02)[6] |
| 4e | Digilab FTS 16 | | | 16K | — | $2 \times$ ca. 4 K | | 1.98 |
| 5a | Polytec MIR 20 | Cooley-Tukey (FFT) | Mathematical phase error correction | 16K (32K) | 1.2M | $2 \times 500$ **K** | $2^n \times 0.6328$ ($-3 \leqq n \leqq +4$) | (0.02)[6] |
| 5b | Polytec MIR 160 | | | | | | | (0.02)[6] |
| 5c | Polytec FIR 30 | Real time | Parabola fit | 4K (16K) | | $2 \times 1.5$ K | 5, 10, 20 | 0.60 |
| 6 | Spectrotherm ST 10 | Cooley-Tukey (FFT) | | 16K (32K) | 128K | $2 \times$ ca. 50 K | 1.2656 | (0.16)[6] |

[1] Capacity of words, word length mostly 16 bit; in brackets: optional.
[2] Which can be stored with the standard version of the instrument.
[3] As limited by the maximum number of data points (single scan, sample and background, widest spectral range, single-sided interferogram).
[4] Analogue computer.
[5] $2 \times .8$ K means 800 data points for sample interferogram, and 800 for background.
[6] Resolution limited at an larger value of $\Delta\tilde{\nu}$ by the maximum path of the movable mirror (cf. Table 3).

The Polytec FIR 30 provides the "parabola fit" and the Coderg FS 4000 a special electronic phase error correction. All instruments with the fast Fourier transform (FFT) correct phase errors in the interferogram mathematically according to a method first proposed by M. Forman [68,70]. This correction procedure was outlined in detail in Section 5.3 (cf. Fig. 41). In addition to Fourier transform and phase error correction, it is advisable to use apodization in Fourier spectroscopy (cf. Sections 2.3 and 3.2). In all commercial instruments, the operator has the choice among a number of different apodization functions.

From Sections 4.3 and 4.4, we recall that either the interferogram sampling points $I(n\Delta s)$ (FFT, Cooley-Tukey-algorithm) or the spectrum points $I(m\Delta\tilde{\nu})$ (Real time) have to be stored in the computer memory. The number of data points to be stored is moderate in the far-infrared and considerably higher in the middle-infrared where it will be of the order of 10,000 to 100,000 for high resolution work. Now, the capacity of the computer core memory ranges from 4K to 16K. That means, 4,000—16,000 data (words in computer language) can be stored there. From this capacity, about one half is needed for the program and the data handling organization. Thus, about 2,000—8,000 data points can be stored in the core memory (cf. Table 4). For high resolution, this is not sufficient. Moreover, one should be able to use multiple scanning and averaging in order to improve the signal-to-noise ratio. And one would like to employ the computer to perform additional calculations with the spectral data, *e.g.* absorbance spectrum evaluated from transmittance. On these grounds, a rather large capacity for storing data is required for the electronic data processing system of a Fourier spectrometer, especially for the middle- and near-infrared. Therefore, most manufactures provide regularly with their instrument a magnetic disk which offers the possibility of storing simultaneously 128,000 to 1,200,000 additional data points. Such a capacity is sufficient even for the highest resolution. And the disk is easily interchangeable. Thus, a multiple of the above numbers is yielded for the limits of this memory. Of course, all manufactures offer optionally further extensions of the data system which may be very helpful when a great number of spectra has to be handled and to be compared with other spectra as will be the case in routine chemical analysis. For the Bruker IFS 115 Fourier spectrometer (No. 2b in Table 4) however, it is advisable to purchase such an optionally extended data system in order to be able to use the full capability of the instrument, even if no extreme data processing is required. For the standard version of all the instruments, an estimate of the maximum number of data points is listed in Table 4. From the total memory, a portion has to be substracted for the programs stored in the computer. As usually half of the available memory is needed for the sample interferogram or spectrum and the other half for the background; the number has been divided by two, *e.g.* $2 \times 4$ K means 4000 sample points and 4000 background points. For the evaluation of the maximum resolution as limited by the electronic data system, we recompile the relations discussed in Section 4.6 and demonstrated in Table 1;

number of interferogram points $N = \frac{s_{\max}}{\Delta s}$

number of spectrum points $M = 2\,\frac{\tilde{\nu}_{\max} - \tilde{\nu}_{\min}}{\Delta\tilde{\nu}}$

where $\Delta s \leqq \frac{1}{2\tilde{\nu}_{max}}$ depending on the potential sampling intervals available at the particular instruments, where $\Delta \tilde{\nu} = \frac{1}{s_{max}}$ (with apodization!), and where it was assumed that two spectrum points are evaluated per resolution width $\Delta \tilde{\nu}$ (cf. Table 1). In case of a He—Ne-laser reference system, the potential sampling intervals are related to the laser wavelength ($\lambda = 0.6238$ $\mu$m) in a simple way (cf. Table 4). Except for one already mentioned instrument (Bruker IFS 115), the maximum resolution all Fourier spectrometers for the middle- and near-infrared is not limited in the standard version by the memory but by the maximum path of the movable mirror (cf. Tables 2—4). For all far-infrared instruments however, the resolution is limited by the data system and not by the mechanics of the interferometer.

It has already been mentioned several times that the electronic computer in a Fourier spectrometer is an unquestionable necessity for performing the Fourier transform of the interferogram in order to obtain the spectrum. But, once there, the computer can be used also for every calculation which is to be executed with interferogram data or spectra. And most manufactures provide for instruments with a sufficiently large core memory (not less than 8 K) a socalled software package which is a collection of helpful additional computer programs. Among these, mostly the following programs are found

a) double precision, *i.e.* doubling the word length from 16 (20) bit to 32 (40) bit. Double precision is necessary for example when the difference of two spectra is required and when a large number of spectra has to be added to obtain a reliable average.

b) arithmetic. This program offers the opportunity to add, substract, multiply or divide any two spectra, *e.g.* to evaluate the %-transmission, the ratio of the sample and of the background spectrum.

c) co-adding of interferograms and averaging of spectra. Theoretically, the Fourier transform is a linear operation. On these grounds, no difference is expected if, for an improvement of the signal-to-noise ratio, the average of a number of interferograms is formed (co-adding) or if this is performed for the spectra.

d) usually the spectra can be converted from percent transmission to absorbance and log-absorbance.

e) for plotting or displaying the spectra, expansions are available for the x- an for the y-scale. One manufacturer (Digilab) also includes a subprogram for three-dimensional plotting.

f) peak finding, integral absorption. This program helps to extract some characteristic data from a spectrum.

g) BASIC and FORTRAN compilers (the FORTRAN compiler only for computers with a core memory not smaller than 16 K). In this case, the customer can write and use programs of his own choice written in BASIC or FORTRAN.

In addition to all these possibilities, the computer is usually employed in controlling and steering the whole instrument.

### 6.5 Final Remarks

In this introduction to Fourier transform spectroscopy, the theory of this spectroscopic method has been set out. Its advantages and diasdvantages have been compared with those of conventional spectroscopy with a diffraction grating. Many practical aspects have been discussed, and it has become clear that Fourier transform spectroscopy is a useful and very powerful tool, especially for energy-limited spectroscopy in the very far-infrared spectral region. Here, its performance cannot be equalled by conventional spectroscopy though it may be reached and exceeded by laser spectroscopy with tunable sources; but this field is beyond the scope of this introduction. Taking Fourier spectroscopy and conventional spectroscopy only, there are cases where comparable final results can be obtained by both methods or where the same results can be obtained without any difficulty by conventional methods [8–10]. Therefore, our view on these methods should not be just "black or white" but more differentiated. In other words, if in a large laboratory a conventional double-beam instrument and a Fourier spectrometer are both available, the spectroscopist should decide on the basis of the problem which instrument to use for an investigation. This introduction is intended to help him in such decisions.

## 7. Appendix 1

For the illustration of the principles of Fourier transform spectroscopy, a spectrum $I(\tilde{\nu})$ has been chosen for which the interferogram $I(s)$ and also the observed spectra for the truncated interferogram can be presented in analytical mathematical form. The spectrum $I(\tilde{\nu})$ is the sum of three "Lorentzian" terms:

$$I(\tilde{\nu}) = \sum_{n=1}^{3} I_n(\tilde{\nu}) = \sum_{n=1}^{3} f_n \frac{\gamma_n^2 \tilde{\nu}_n^2}{(\tilde{\nu}_n^2 - \tilde{\nu}^2)^2 + \gamma_n^2 \tilde{\nu}^2} \tag{A 1.1}$$

where $\tilde{\nu}_n$, $\gamma_n$ and $f_n$ are the characteristic frequency, the width and an intensity factor of the Lorentzian terms, respectively. The interferogram corresponding to $I(\tilde{\nu})$ is the sum of three damped cosine waves

$$I(s) = \sum_{n=1}^{3} \pi \gamma_n f_n \left[1 + \frac{\tilde{\nu}_n}{\tilde{\nu}_n'} \cos(2\pi \tilde{\nu}_n' |s| - \varphi_n) \mathrm{e}^{-\pi \gamma_n |s|}\right]$$

or

$$\bar{I}(s) = \sum_{n=1}^{3} \pi \gamma_n f_n \frac{\tilde{\nu}_n}{\tilde{\nu}_n'} \cos(2\pi \tilde{\nu}_n' |s| - \varphi_n) \mathrm{e}^{-\pi \gamma_n |s|} \tag{A 1.2}$$

where

$$\varphi_n = \text{are} \sin \frac{\gamma_n}{2\tilde{\nu}_n} \quad \text{and} \quad \tilde{\nu}_n' = \tilde{\nu}_n \cos \varphi_n = \sqrt{\tilde{\nu}_n^2 - \frac{1}{4}\gamma_n^2}$$

For $s = 0$ and for $s \to \infty$, the interferogram given by Eq. (A 1.2) exhibits the properties quoted in Section 3.2. The functions $I(\tilde{\nu})$ and $\bar{I}(s)$ are a Fourier transform pair well known from the theory of forced vibrations and resonance of damped oscillators. One reason why they are very useful to demonstrate problems of Fourier transform spectroscopy is that the different contributions to $I(\tilde{\nu})$ and to $\bar{I}(s)$ can be studied separately (cf. Fig. 12).

The Fourier transform of a finite interferogram without apodization is represented by the following expression:

$$I_{obs}(\tilde{\nu}) = \sum_{n=1}^{3} I_n(\tilde{\nu}) \left[1 - F_{1n} \cos(2\pi\tilde{\nu}s_{max}) + F_{2n} \frac{\tilde{\nu}}{\tilde{\nu}_n} \sin(2\pi\tilde{\nu}s_{max})\right]$$

with (A 1.3)

$$F_{1n} = \frac{1}{\pi\gamma_n f_n} \bar{I}_n(s_{max}) + \frac{\tilde{\nu}^2 - \tilde{\nu}_n^2}{\gamma_n \tilde{\nu}_n'} \sin(2\pi\tilde{\nu}_n' s_{max}) e^{-\pi\gamma_n s_{max}}$$

$$F_{2n} = \frac{\tilde{\nu}^2 - \tilde{\nu}_n^2 - \gamma_n^2}{\pi\gamma_n f_n} \bar{I}_n(s_{max}) - \frac{\tilde{\nu}_n}{\tilde{\nu}_n'} \sin(2\pi\tilde{\nu}_n' s_{max}) e^{-\pi\gamma_n s_{max}} .$$

The function $I_{obs}(\tilde{\nu})$ can easily be evaluated for any value of $\tilde{\nu}$. Further, the analytical properties may be considered. Obviously, the asymptotic value of $I_{obs}(\tilde{\nu})$ for $s_{max} \to \infty$ is the true spectrum $I(\tilde{\nu})$. On the other hand, if the widths of the Lorentzian terms are reduced drastically, we obtain

$$\sum_{n=1}^{3} \lim_{\gamma_n \to 0} \frac{1}{\gamma_n} I_n(\tilde{\nu}) \left[1 - F_{1n} \cos(2\pi\tilde{\nu}s_{max}) + F_{2n} \frac{\tilde{\nu}}{\tilde{\nu}_n} \sin(2\pi\tilde{\nu}s_{max})\right]$$

$$= \frac{1}{2} \sum_{n=1}^{3} f_n \left[\frac{\sin[2\pi(\tilde{\nu} - \tilde{\nu}_n) s_{max}]}{[\tilde{\nu} - \tilde{\nu}_n]} + \frac{\sin[2\pi(\tilde{\nu} + \tilde{\nu}_n) s_{max}]}{[\tilde{\nu} + \tilde{\nu}_n]}\right] \qquad \text{(A 1.4)}$$

This result is what we expect for three narrow lines [cf. Eq. (2.21)]. The analytic expression for the observed spectrum $I_{obs}(\tilde{\nu})$ in the case of triangular apodization was also derived, but it is too lengthy to be reproduced here. It has, however, been applied to derive the corresponding graphs in Fig. 11.

## 8. Appendix 2

This short derivation of the Cooley-Tukey algorithm is given to demonstrate the principle; it cannot illuminate all the details and aspects of an actual computation. In that respect, the reader is referred to the literature [49,50,97].

The conventional computation of the Fourier transform with a digital computer is done according to Eq. (4.8) in Section 4. For our purpose, it is convenient to proceed to a complex notation

$$I_{obs}(m\,\Delta\tilde{\nu}) = \sum_{n=-N}^{N-1} \bar{I}(n\,\Delta s)\, e^{i\pi\frac{m \cdot n}{N}} \qquad \text{(A 2.1)}$$

with $0 \leqq m \leqq N-1$ and where $\bar{I}(n\Delta s) = S(n\Delta s)\cdot \bar{I}(n\Delta s)\cdot \Delta s$ for all $n \geqq 0$ and $\bar{I}(-n\Delta s) = \bar{I}(n\Delta s)$ for all $n<0$. In order to restrict the sum to $2N$ terms, the term with $n=N$ was omitted. This can be done without change of $I_{\text{obs}}$ since the apodization function has the property $S(N\Delta s) = S(s_{\max}) = 0$. Now, it is useful to get rid of the negative values of $n$ in Eq. (A 2.1). We substitute $n+2N$ for $n<0$ and put, consequently,

$$\bar{I}([n+2N]\Delta s) = \bar{I}(-n\Delta s)\,.$$

This yields from Eq. (A 2.1) remembering that $e^{i\pi\frac{m\,2N}{N}} = e^{2\pi i m} = 1$

$$I_{\text{obs}}(m\Delta\tilde{\nu}) = \sum_{n=0}^{2N-1} \bar{I}(n\Delta s)\, e^{i\pi\frac{m\,n}{N}}\,. \tag{A 2.2}$$

The derivation of the Cooley-Tukey algorithm now requires that all integers $m$, $n$, $N$ be converted to binary numbers to take advantage of the periodicity of the function $e^{i\varphi}$. Let us assume

$$N = 2^k \quad \text{or} \quad 2N = 2^{k+1}\,. \tag{A 2.3}$$

If Eq. (A 2.3) is not fulfilled for an interferogram determined experimentally with a certain number of data, it is advisable to add the required number of data points with $\bar{I}(n\Delta s) = 0$. For $m$ and $n$, the conversion to binary numbers is

$$m = \sum_{\mu=0}^{k} m_\mu 2^\mu \quad \text{and} \quad n = \sum_{\lambda=0}^{k} n_\lambda 2^\lambda\,, \tag{A 2.4}$$

where all the $m_\mu$ and $n_\lambda$ may have only the values zero or one. Since $m$ is restricted to the range $0 \leqq m \leqq N-1 = 2^k-1$, the coefficient $m_k$ of $2^k$ is zero in all cases. An example of conversion to binary numbers is

$$\begin{aligned}107 &= 1\cdot 10^2 + 0\cdot 10^1 + 7\cdot 10^0\\ &= 1\cdot 64 + 1\cdot 32 + 0\cdot 16 + 1\cdot 8 + 0\cdot 4 + 1\cdot 2 + 1\cdot 1\\ &= 1\cdot 2^6 + 1\cdot 2^5 + 0\cdot 2^4 + 1\cdot 2^3 + 0\cdot 1^2 + 1\cdot 2^1 + 1\cdot 2^0.\end{aligned}$$

The conversion to binary numbers means in effect that a number is not expressed in terms of powers of ten, as in the decimal system, but in terms of powers of two.

With the relation (A 2.4) we obtain

$$e^{i\pi\frac{m\cdot n}{N}} = \prod_{\lambda=0}^{k} \exp\left\{-\,2\pi i n_\lambda \left[\sum_{\mu=0}^{k-\lambda} m_\mu\, 2^{\mu+\lambda-k-1}\right]\right\}. \tag{A 2.5}$$

In Eq. (A 2.5), all integral multiples of $2\pi$ in the exponent have been removed. The spectrum and the interferogram data are redefined in terms of the binary bits $m_\mu$ and $n_\lambda$:

$$\begin{aligned} I_{\text{obs}}(m\Delta\tilde{\nu}) &= I(m_k, m_{k-1}, \ldots, m_1, m_0) \\ \tilde{I}(n\Delta s) &= \hat{I}(n_k, n_{k-1}, \ldots, n_1, n_0)\,. \end{aligned} \tag{A 2.6}$$

The Fourier transform [see Eq. (A 2.2)] then reads

$$\begin{aligned} I(m_k,\ldots,m_0) = &\sum_{n_0=0}^{1} \exp\left\{-2\pi i n_0 \sum_{\mu=0}^{k} m_\mu\, 2^{\mu-k-1}\right\} \cdot \sum_{n_1=0}^{1} \exp\left\{-2\pi n_1 \sum_{\mu=0}^{k-1} m_\mu\, 2^{\mu-k}\right\} \\ &\ldots \sum_{n_{k-1}=0}^{1} \exp\left\{-2\pi i n_{k-1}\left[\frac{m_0}{4}+\frac{m_1}{2}\right]\right\} \\ &\cdot \sum_{n_k=0}^{1} \exp\left\{-2\pi i n_k \frac{m_0}{2}\right\} \cdot \hat{I}(n_k,\ldots,n_0)\,. \end{aligned} \tag{A 2.7}$$

This is the basic formula for the Cooley-Tukey algorithm. But in order to reach some understanding of its advantages, we have to consider the procedure of computing $I(m_k,\ldots,m_0)$ in separate steps.

At the beginning, the memory of the computer contains the interferogram (see Fig. 51). Before starting the computation it is convenient to invert the order of the binary bits of these data:

$$\hat{I}(n_k, n_{k-1},\ldots, n_1, n_0) \rightarrow \hat{I}(n_0, n_1,\ldots, n_{k-1}, n_k) \tag{A 2.8}$$

The first step of the Fourier transform in this algorithm is the summation over $n_k$ [see Eq. (A 2.7) and Fig. 51]:

$$\begin{aligned} S_1(n_0,n_1,\ldots n_{k-1}|m_0) &= \sum_{n_k=0}^{1} \exp\left\{-2\pi i n_k \frac{m_0}{2}\right\} \cdot \hat{I}(n_0,\ldots,n_k) \\ &= \hat{I}(n_0,\ldots,n_{k-1},0) + \hat{I}(n_0,\ldots,n_{k-1},1)\,e^{-\pi i m_0}\,. \end{aligned} \tag{A 2.9}$$

Since $n_k$ has been replaced by $m_0$, the quantities $S_1$ need the same space in the memory of the computer as $\hat{I}(n_0,\ldots n_k)$ did (see Fig. 51). In the following steps, the summation is carried out successively over $n_{k-1}$, $n_{k-2}$, and so on according to

$$\begin{aligned} S_2(n_0,\ldots,n_{k-2}|m_1,m_0) &= \sum_{n_{k-1}=0}^{1} \exp\left\{-2\pi i n_{k-1}\left(\frac{m_0}{4}+\frac{m_1}{2}\right)\right\} S_1(n_0,\ldots,n_{k-1}|m_0) \\ &= S_1(n_0,\ldots,n_{k-2},0|m_0) + S_1(n_0,\ldots,n_{k-2},1|m_0)\,e^{-\pi i\left(m_1+\frac{m_0}{2}\right)} \end{aligned} \tag{A 2.10}$$

Location no. / Initial contents

| Decimal $n$ | Binary $n_2 n_1 n_0$ | Initial contents |
|---|---|---|
| 0 | 0 0 0 | $\tilde I(0)$ |
| 1 | 0 0 1 | $\tilde I(1)$ |
| 2 | 0 1 0 | $\tilde I(2)$ |
| 3 | 0 1 1 | $\tilde I(3)$ |
| 4 | 1 0 0 | $\tilde I(4)$ |
| 5 | 1 0 1 | $\tilde I(5) = \tilde I(-3)$ |
| 6 | 1 1 0 | $\tilde I(6) = \tilde I(-2)$ |
| 7 | 1 1 1 | $\tilde I(7) = \tilde I(-1)$ |

→ Binary bit inversion

| $n_0 n_1 n_2$ | |
|---|---|
| 0 0 0 | $\tilde I(0)$ |
| 0 0 1 | $\tilde I(4)$ |
| 0 1 0 | $\tilde I(2)$ |
| 0 1 1 | $\tilde I(6)$ |
| 1 0 0 | $\tilde I(1)$ |
| 1 0 1 | $\tilde I(5)$ |
| 1 1 0 | $\tilde I(3)$ |
| 1 1 1 | $\tilde I(7)$ |

→ *1st* step

| $n_0 n_1 m_0$ | |
|---|---|
| 0 0 0 | $\tilde I(0) + \tilde I(4)$ |
| 0 0 1 | $\tilde I(0) - \tilde I(4)$ |
| 0 1 0 | $\tilde I(2) + \tilde I(6)$ |
| 0 1 1 | $\tilde I(2) - \tilde I(6)$ |
| 1 0 0 | $\tilde I(1) + \tilde I(5)$ |
| 1 0 1 | $\tilde I(1) - \tilde I(5)$ |
| 1 1 0 | $\tilde I(3) + \tilde I(7)$ |
| 1 1 1 | $\tilde I(3) - \tilde I(7)$ |

→ *2nd* step

| $n_0 m_1 m_0$ | |
|---|---|
| 0 0 0 | $[\tilde I(0) + \tilde I(4)] + [\tilde I(2) + \tilde I(6)]\,e^{-i0}$ |
| 0 0 1 | $[\tilde I(0) - \tilde I(4)] + [\tilde I(2) - \tilde I(6)]\,e^{-i\frac{\pi}{2}}$ |
| 0 1 0 | $[\tilde I(0) + \tilde I(4)] + [\tilde I(2) + \tilde I(6)]\,e^{-i\pi}$ |
| 0 1 1 | $[\tilde I(0) - \tilde I(4)] + [\tilde I(2) - \tilde I(6)]\,e^{-i\frac{3\pi}{2}}$ |
| 1 0 0 | $[\tilde I(1) + \tilde I(5)] + [\tilde I(3) + \tilde I(7)]\,e^{-i0}$ |
| 1 0 1 | $[\tilde I(1) - \tilde I(5)] + [\tilde I(3) - \tilde I(7)]\,e^{-i\frac{\pi}{2}}$ |
| 1 1 0 | $[\tilde I(1) + \tilde I(5)] + [\tilde I(3) + \tilde I(7)]\,e^{-i\pi}$ |
| 1 1 1 | $[\tilde I(1) - \tilde I(5)] + [\tilde I(3) - \tilde I(7)]\,e^{-i\frac{3\pi}{2}}$ |

→ *3rd* and final step (for $N = M = 4$)

| $m_2 m_1 m_0$ | |
|---|---|
| 0 0 0 | $\langle[\tilde I(0) + \tilde I(4)] + [\tilde I(2) + \tilde I(6)]\,e^{-i0}\rangle + \langle[\tilde I(1) + \tilde I(5)] + [\tilde I(3) + \tilde I(7)]\,e^{-i0}\rangle\ e^{-i0} = \sum_n \tilde I(n)\,e^{-i\pi\frac{n\cdot 0}{4}} = I(0\Delta\tilde\nu)$ |
| 0 0 1 | $\langle[\tilde I(0) - \tilde I(4)] + [\tilde I(2) - \tilde I(6)]\,e^{-i\frac{\pi}{2}}\rangle + \langle[\tilde I(1) - \tilde I(5)] + [\tilde I(3) - \tilde I(7)]\,e^{-i\frac{\pi}{2}}\rangle\ e^{-i\frac{\pi}{4}} = \sum_n \tilde I(n)\,e^{-i\pi\frac{n\cdot 1}{4}} = I(1\cdot\Delta\tilde\nu)$ |
| 0 1 0 | $\langle[\tilde I(0) + \tilde I(4)] + [\tilde I(2) + \tilde I(6)]\,e^{-i\pi}\rangle + \langle[\tilde I(1) + \tilde I(5)] + [\tilde I(3) + \tilde I(7)]\,e^{-i\pi}\rangle\ e^{-i\frac{\pi}{2}} = \sum_n \tilde I(n)\,e^{-i\pi\frac{n\cdot 2}{4}} = I(2\cdot\Delta\tilde\nu)$ |
| 0 1 1 | $\langle[\tilde I(0) - \tilde I(4)] + [\tilde I(2) - \tilde I(6)]\,e^{-i\frac{3\pi}{2}}\rangle + \langle[\tilde I(1) - \tilde I(5)] + [\tilde I(3) - \tilde I(7)]\,e^{-i\frac{3\pi}{2}}\rangle\ e^{-i\frac{3\pi}{4}} = \sum_n \tilde I(n)\,e^{-i\pi\frac{n\cdot 3}{4}} = I(3\cdot\Delta\tilde\nu)$ |

Fig. 51. Illustration of the Cooley-Tukey algorithm for $N = 4$

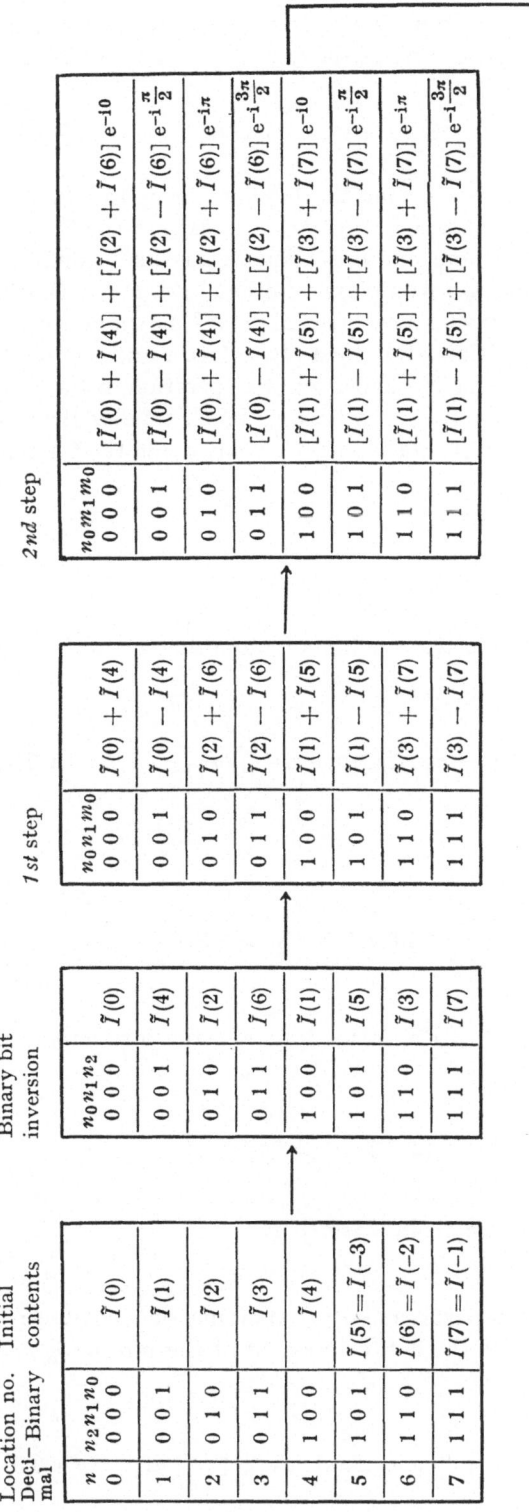

and similar expressions for $S_3$, $S_4$, etc. From Eq. (A 2.7) it is obvious that there are $(k+1)$ steps until the computation arrives at the final result (see Fig. 51):

$$I(m_k, m_{k-1}, \ldots, m_0) = S_{k+1}(m_k, m_{k-1}, \ldots, m_0) \tag{A 2.11}$$

The main advantage of this method is that the required computer time is proportional to $(k+1)\,N = N\,({}_2\log N + 1) = N\,{}_2\log 2N$. In each of the $(k+1)$ steps $N$ operations are performed, where one operation means one (complex) multiplication by a phase factor and two (complex) additions. For conventional integration, the required computer time is proportional to $N^2$ as there are $N^2$ operations consisting of one (real) multiplication by a phase factor and one (real) addition. Although the computation time is smaller for one operation of the conventional method, the factor ${}_2\log 2N = k+1$ will be so much smaller than $N$ for large $N$ that a considerable amount of time is saved by employing the Cooley-Tukey algorithm.

## 9. Appendix 3

For phase modulation, the interferogram is a function of path difference $s$ and the modulation $\sigma(t)$ [cf. Eq. (4.9)]:

$$I_\varphi(s,t) = 2\int_0^\infty I(\tilde{\nu})\,\{1 + \cos(2\pi\tilde{\nu}[s + \sigma(t)])\}\,d\tilde{\nu} \tag{A 3.1}$$

where, for square-wave modulation,

$$\sigma(t) \begin{cases} +\sigma_0 & \text{for } 0 < t < \frac{1}{2}T_0 \\ -\sigma_0 & \text{for } \frac{1}{2}T_0 < t < T_0 \end{cases} \quad \text{and } \sigma(t + T_0) = \sigma(t)\,.$$

According to the addition theorem of the cosine function, the interferogram may be separated into two parts

$$\begin{aligned} I_\varphi(s,t) &= 2\int_0^\infty I(\tilde{\nu})\,\{1 + \cos(2\pi\tilde{\nu}s)\cos(2\pi\tilde{\nu}\sigma[t])\}\,d\tilde{\nu} \\ &\quad - 2\int_0^\infty I(\tilde{\nu})\sin(2\pi\tilde{\nu}s)\sin(2\pi\tilde{\nu}\sigma[t])\,d\tilde{\nu}\,. \end{aligned} \tag{A 3.2}$$

Inserting $\sigma(t)$, we learn that the first part is independent of time since $\cos(2\pi\tilde{\nu}\sigma_0) = \cos(-2\pi\tilde{\nu}\sigma_0)$. Thus, the first part is the d.c. component of the interferogram signal

$$\text{d.c. component} = 2\int_0^\infty I(\tilde{\nu})\,\{1 + \cos(2\pi\tilde{\nu}s)\cos(2\pi\tilde{\nu}\sigma_0)\}\,d\tilde{\nu} \tag{A 3.3}$$

The sine function, on the other hand, is an odd function so that the second part is time-dependent and contains the a.c. components. For these, it is useful to introduce the Fourier series with the basic frequency $f_0 = \frac{1}{T_0}$ and the harmonics $nf_0$. In this way, we obtain

$$\text{a.c. components} = -2\int_0^\infty I(\tilde{\nu}) \sin(2\pi\tilde{\nu}s)\,[\pm \sin(2\pi\tilde{\nu}\sigma_0)]\,d\tilde{\nu}$$

$$= \frac{4}{\pi}\left\{-2\int_0^\infty \sin(2\pi\tilde{\nu}s)\sin(2\pi\tilde{\nu}\sigma_0)\,d\tilde{\nu}\right\}\cdot\sum_{\mu=0}^{\infty}\frac{\sin(2\pi[2\mu+1]f_0 t)}{[2\mu+1]}. \tag{A 3.4}$$

Only the component with $f_0$ [$\sim \sin(2\pi f_0 t)$] is transmitted by the narrow-band amplifier and the phase-sensitive rectifier [see Eq. (4.11)] and all the higher harmonics are rejected. Thus, the interferogram signal obtained at the low-pass filter is proportional to the factor of this frequency component (curly brackets).

Omitting the factor $4/\pi$, which is not important in our context, we can write the interferogram signal as in Eq. (4.12).

## 10. References

A) Books and review articles dealing with Fourier transform spectroscopy and in some cases comparing it with other methods:

1) Stewart, J. E.: Infrared spectroscopy, New York: Dekker 1970.
2) Chantry, G. W.: Submillimetre spectroscopy. London: Academic Press 1971.
3) Möller, K. D., Rothschild, G.: Far infrared spectroscopy. New York: Wiley Interscience 1971.
4) Bell, R. J.: Introductory fourier transform spectroscopy. New York: Academic Press 1972.
5) Richards, P. L.: J. Opt. Soc. Am. *54*, 1474 (1964).
6) Loewenstein, E. V.: Appl. Opt. *5*, 845 (1966).
7) Vanasse, G. A., Sakai, H.: Progress in optics, Vol. VI, p. 261. Amsterdam: North Holland 1967.
8) A critical review of Fourier spectroscopy and a comparison with other methods is presented in
Jacquinot, P.: Appl. Optics *8*, 497 (1967).
Gebbie, H. A.: Appl. Optics *8*, 501 (1967).
Kneubühl, F.: Appl. Optics *8*, 505 (1967).
Mertz, L.: Appl. Optics *10*, 386 (1971).
9) Connes, P.: "How light is analyzed", Scientific American *219*, 72 (1968).
10) Genzel, L: Fourierspektroskopie, Plenarvorträge der 33. Physikertagung 1968 in Karlsruhe, p. 128, B. G. Teubner.
Genzel, L.: Z. Anal. Chem. *273*, 91 (1975).
11) Geick, R.: Chem. Labor Betrieb *23*, 194, 250 und 300 (1972)
Geick, R.: Meßtechnik *2*, 43 (1974).
12) Ziessow, D.: On-line Rechner in der Chemie, Berlin: de Gruyter 1973.

B) Papers containing bibliographies on the far infrared:

13) Palik, E. D.: J. Opt. Soc. Am. *50*, 1329 (1960); (covers the period from 1892 to 1960, an extension to 1969 is published in Ref. 3).
14) Bloor, D.: Infrared Phys. *10*, 1, (1970); (covers the period from 1920 to 1969).

C) Fourier transforms:

15) Champeney, D. C.: Fourier transforms and their physical applications. London: Academic Press 1973.

D) Some historical papers:

16) Fizeau, H.: Ann. Chim. Phys. (3) *66*, 429 (1862).
17) Michelson, A. A.: Phil. Mag. (5) *31*, 256 (1891).
18) Michelson, A. A.: Phil. Mag. (5) *34*, 280 (1892).
19) Rubens, H., Wood, R. W.: Phil. Mag. *21*, 249 (1911).
20) Jacquinot, P., Dufour, C. J.: J. Rech. C. N. R. S. *6*, 91 (1948).
21) Fellgett, F. S.: Thesis Univ. of Cambridge, 1951.
22) Strong, J.: J. Opt. Soc. Am. *44*, 352 (1954).
23) Gebbie, H. A., Vanasse, G. A.: Nature *178*, 432 (1956).
24) Mertz, L.: J. Opt. Soc. Am. *46*, 548 (1956).
25) Strong, J. D.: J. Opt. Soc. Am. *47*, 354 (1957).
26) Genzel, L., Weber, R.: Z. Angew. Phys. *10*, 127 und 195. (1958).
27) Strong, J. D., Vanasse, G.: J. Opt. Soc. Am. *49*, 844 (1959).
28) Genzel, L.: J. Mol. Spectr. *4*, 261 (1960).

E) Further references to the Introduction:

29) See, for example, Ref. [2], Chapters 1 and 2.
30) See, for example, the instruction manuals of modern commercial Fourier spectrometers.
31) Milward, R. C.: The Spex Speaker, Vol. XVII, 1 (1972).

F) Special references to Section 2, "Fundamentals of spectroscopy" for further information, see any book on optics or the theory of electrodynamics, *e.g.*:

32) Strong, J.: Concepts of classical optics. San Francisco: Freeman 1958. — Born, M., Wolf, E.: Principles of optics. London: Pergamon, 1964. — Hund, F.: Theoretische Physik, Vol. 2. Stuttgart: B. G. Teubner, 1957. — Lüscher, E.: Experimentalphysik, Vol. II, B—J Hochschultaschenbücher Nr. 115/115a.
33) For further details on grating spectroscopy see Ref. [2], Chapter 2, and Ref. [3], Chapter 1.
34) The mathematical proofs are to be found in Ref. [4] Chapter 3.
35) In the literature, the following abbreviations are sometimes used:

$$\frac{\sin \pi x}{\pi x} = \mathrm{sinc}\,(x) = \mathrm{dif}\,(x) \quad \text{and} \quad \left(\frac{\sin \pi x}{\pi x}\right)^2 = \mathrm{sinc}^2\,(x) = \mathrm{dif}^2\,(x).$$

36) Happ, H., Genzel, L.: Infrared Phys. *1*, 39 (1961).

G) Special references to Section 3, "Spectroscopy":

37) Czerny, M., Turner, A.: Z. Physik *61*, 792 (1930).
38) Genzel, L., Eckhardt, W.: Z. Physik *139*, 578 (1954).
39) Moser, J. F., Steffen, H., Kneubühl, F. K.: Helv. Phys. Acta *41*, 607 (1968).
40) Geick, R.: Z. Physik *163*, 499 (1963).
41) Krüger, R. A., Anderson, L. W., Roesler, F. L.: Appl. Opt. *12*, 533 (1973) and J. Opt. Soc. Am. *62*, 938 (1972).

H) Special references to Section 4, "Practice of Fourier transform spectroscopy":

42) Perry, C. H., Geick, R., Young, E. F.: Appl. Opt. *5*, 1171 (1966).
43) Hanel, R., Forman, M., Meilleur, T., Westcott, R., Pritchard, J.: Appl. Opt. *8*, 2059 (1969).
44) Richards, P. L.: J. Opt. Soc. Am. *54*, 1474 (1964).
45) Martin, D. H., Puplett, E.: Infrared Phys. *10*, 105 (1970).
46) Mager, H. J.: Infrared Phys. *13*, 7 (1973).
47) Burrough, W. J., Chamberlain, J.: Infrared Phys. *11*, 1 (1971).
48) Ridyard, J. N. A.: J. Physique Radium *28*, C 2—62 (1962).
49) Cooley, J. W., Tukey, J. W.: Math. Computation *19*, 296 (1965).
50) Forman, M. L.: J. Opt. Soc. Am. *56*, 978 (1966).

51) Yoshinaga, H., Fujita, S., Minami, S., Suemoto, Y., Jnoue, M., Chiba, K., Nakano, K., Yoshida, S., Sugimori, H.: Appl. Opt. *5*, 1159 (1966).
52) Hoffmann, J. E., Jr.: Appl. Opt. *8*, 323 (1969).
53) Milward, R. C., Irslinger, C., Lossau, H.: Meßtechnik *79*, 1 (1971).
54) Chamberlain, J.: Infrared Phys. *11*, 25 (1971).
55) Chamberlain, J., Gebbie, H. A.: Infrared Phys. *11*, 57 (1971).
56) Birch, J. R.: Infrared Phys. *12*, 29 (1972).
57a) Fleming, J. W., Chamberlain, J.: Infrared Phys. *14*, 277 (1974).
57b) Murphy, R. E., Cook, F. H., Sakai, H.: J. Opt. Soc. Am. *65*, 600 (1975).
57c) Koenig, J. L., Tabb, D. L.: Digilab application note No. 18, March 1975.
58) Connes, J., Connes, P.: J. Opt. Soc. Am. *56*, 896 (1966).
59a) Bell, E. E.: Infrared Phys. *6*, 57 (1966).
59b) Russel, E. E., Bell, E. E.: Infrared Phys. *6*, 75 (1966).
60a) Gast, J., Genzel, L.: Optics Communications *8*, 26 (1973).
60b) Gast, J., Genzel, L., Zwick, U.: IEEE Transactions on microwave theory and techniques, Vol. MTT 22, 1026 (1974).
61) Parker, T. J., Chambers, W. G., Angress, J. F.: Infrared Phys. *14*, 207 (1974).
62a) Sanderson, R. B.: Appl. Opt. *6*, 1527 (1967).
62b) Chamberlain, J., Gibbs, J. E., Gebbie, H. A.: Infrared Phys. *9*, 185 (1969).

J) Special references to Section 5, "Advantages and Disadvantages":

63) Jacquinot, P.: Rep. Prog. Phys. *23*, 267 (1960).
64) Fellgett, P.: J. Phys. Radium *19*, 187 (1958).
65) Connes, P.: Aspen Int. Conf. on Fourier Spectroscopy 1970, p. 121 (G. A. Vanasse, A. J. Stair and D. J. Baker, eds.) AFCRL (Air Force Cambridge Research Laboratories, Hanscomb Field, Cambr., Massachusetts) -71-0019, 5. Jan. 1971, Spec. Rep. No. 114
66) Detailed discussions are to be found in Refs. 4) and 11).
67) See, for example, Ref. 4), pp. 133—168.
68) Forman, M. L., Steel, W. H., Vanasse, G. A.: J. Opt. Soc. Am. *56*, 59 (1966).
69) Mertz, L.: Infrared Phys. *7*, 17 (1967).
70) Sakai, H., Vanasse, G. A., Forman, M. L.: J. Opt. Soc. Am. *58*, 84 (1968).
71) Bohdanski, J.: Z. Physik *149*, 383 (1957).
72) Ibbet, R. N., Aspinal, D., Grainger, J. F.: Appl. Optics *7*, 1089, (1968).
73) Decker, J. A., Harwit, M. O.: Appl. Optics *7*, 2205 (1968).
74) Sloane, N. J. A., Fine, T., Phillips, P. G., Harwit, M. O.: Appl. Optics *8*, 2103 (1969).
75) Nelson, E. D., Fredman, M. L.: J. Opt. Soc. Am. *60*, 1664 (1970).
76) Decker, J. A.: Appl. Optics *10*, 1971 (1971).
77) Decker, J. A.: J. Anal. Chemistry *44*, 127A (1972).
78) Hartwit, M., Phillips, P. G., Fine, T., Sloane, N. J. A.: Appl. Optics *9*, 1149 (1970).
79) Hirschfeld, T., Wyntjes, G.: Appl. Optics *12*, 2877 (1973).
80) Hansen, P., Strong, J.: Appl. Optics *11*, 502 (1972).
81) Phillips, P. G., Briotta, D. A.: Appl. Optics *13*, 2233 (1974).
82) Hanel, R. A., Schlachman, B., Rogers, D., Vanous, D.: Appl. Optics *10*, 1376 (1971).
83) Hanel, R., Schlachman, B., Breihan, E., Bywater, R., Chapman, F., Rhodes, M., Rogers, D., Vanous, D.: Appl. Optics *11*, 2625 (1972).
84) Beckman, J. E., Shaw, J. A.: Infrared Physics *14*, 61 (1974).
85) Beckman, J. E., Harries, J. E.: Appl. Optics *14*, 471 (1975).
86) Sheahen, T. P.: Appl. Optics *13*, 2907 (1974).
87) Kunz, L. W., Goorvitch, D.: Appl. Optics *13*, 1077 (1974).
88) Goorvitch, D.: Appl. Optics *14*, 1387 (1975).
89) Sanderson, R. B., Bell, E. E.: Appl. Optics *12*, 266 (1973).
90) Sheahen, T. P.: Appl. Spectroscopy *28*, 283 (1974).
91) Sheahen, T. P.: J. Opt. Soc. Am. *64*, 485 (1974).
92) Sheahen, T. P.: Appl. Optics *14*, 1004 (1975).
93) Klug, D. D., Whalley, E.: J. Opt. Soc. *64*, 1019 (1974).
94) Filler, A. S.: J. Opt. Soc. America *63*, 589 (1973).
95) Tsoi, V. I., Sokolova, T. N.: Optics and Spectroscopy *36*, 240 (1974).

[96] Bell, E. E., Sanderson, R. B.: Appl. Optics *11*, 688 (1972).
[97] Ziegler, H.: Infrared Physics *15*, 19 (1975).
[98] Knözinger, E.: Berichte der Bunsen-Gesellschaft *78*, 1199 (1974).
[99] Ahern, F. J., Pritchet, C.: Appl. Optics *13*, 2240 (1974).
[100] Vogel, P., Genzel, L.: Infrared Phys. *4*, 257 (1964).

Received March 18, 1974, July 21, 1975